I0036660

Current Trends on Glass and Ceramic Materials

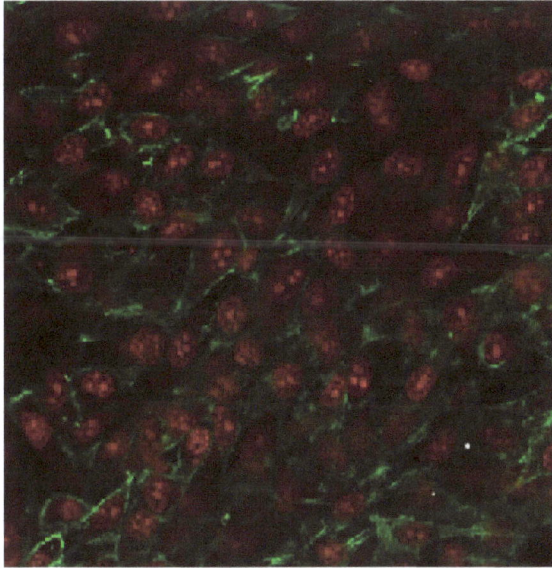

Edited by

Sooraj H. Nandyala
FCUP/INESC TEC
University of Porto, Porto
Portugal

&

José D. Santos

FEUP, University of Porto, Porto
Portugal

Caption: MG63 osteoblastic-like cell culture established over the surface of hydroxyapatite sample for 4 days. Green staining corresponds to the F-actin fibers (stained with phalloidin-conjugated Alexa Fluor 488), while red straining corresponds to the nucleus (stained with propidium iodine). 400x magnification.

CONTENTS

CHAPTERS

FOREWORD BY C.N.R. RAO

The accidental invention of glass-ceramic materials in the mid 1950s is due to the well-known glass chemist, S. D. Stookey. However, it was Larry Hench's Bioglass® that was the first synthetic material specifically developed to form a chemical bond with bones, for regeneration applications in the late 1960s. Since then many researchers have developed a variety of novel materials for biomedical applications. The performance of biomaterials depends upon chemical and biochemical interactions between the biomaterials and their hosts. Several synthetic materials have proven to be of clinical use, enhancing the process of periodontal regeneration. Among them, Bonelike®, a glass reinforced hydroxyapatite has been implanted in intra-bony periodontal defects and has demonstrated increased bone regeneration, as assessed by clinical and radiological parameters.

There has also been rapid development in rare-earth containing materials for optical applications. More recently, hydroxyapatite substituted with rare-earth ions has become the focus of interest for bone related applications.

This eBook entitled "Current Trends on Glass Ceramics Materials" should be useful to materials scientists, engineers, as well as surgeons, and to provide an insight into the current developments in the field. It should also be useful to research scientists pursuing research on new generation biomaterials. I recommend the eBook for wide use by all those concerned with ceramic biomaterials.

C.N.R. Rao
National Research Professor
Honorary President & Linus Pauling Research Professor
Jawaharlal Nehru Centre for Advanced Scientific Research
Bangalore - India

FOREWORD BY GINTARAS JUODŽBALYS

Human skeletal bone loss is a major health concern in the twenty first century, with massive socioeconomic implications. Bone loss can appear because of traumatic injury, bone cancer or birth defects. Furthermore there is a significant proportion of patients with bony defects or non-unions that are not amenable to healing only by direct fixation. Unfortunately, autograft, which is a current gold standard in bone grafts, has some serious shortcomings. These are related to the need for extra surgical procedures, with their associated blood loss, postoperative pain and possible complications, such as injury to nearby nerves or major blood vessels. Also, the amount of bone tissue that can be harvested is limited.

Nowadays alloplasts, synthetically derived bone graft biomaterials are replacing autograft in craniofacial and orthopedic surgery. They are becoming widely used because of their promising results as bone regenerative materials. Every day, people in different countries receive bone biomaterials for replacement of diseased tissue and in the therapy for non-union defects. Consequently, there is an increasing demand for orthopedic implants that integrate into recipient's body tissue and as fillers to treat critical size defects.

Interdisciplinary studies performed by materials scientists, engineers as well as by surgeons, showed that glass and ceramic materials are promising biomaterials that can be successfully used for bone regeneration. Furthermore, the development of bio-glass, glass reinforced hydroxyapatite and other new biocompatible materials are promising to enhance cell adhesion and proliferation, which is extremely important in bone tissue engineering.

This eBook is intended to demonstrate how materials in the field of glass and ceramics can be employed for different applications.

Gintaras Juodžbalys
President of the Baltic Osseoingeration Academy &
Implantologist and Maxillofacial Surgeon
Lithuanian University of Health Sciences
Kaunas, Lithuania

PREFACE

Glass and ceramics have generated a great deal of interest as potential materials for biomedical and technological applications. The aim of the eBook entitled *"Current Trends on Glass and Ceramic Materials"* is to extensively review the latest developments in glass and ceramic materials. Bio-glass and glass reinforced hydroxyapatite related research and its novel applications are essential for the health sector. In fact, a wide range of investigation is on-going worldwide by eminent scholars in order to further develop innovative materials for the next generation of applications. Therefore, this eBook is intended as a reference guide for all the scientific community and concurrently, the authors hope that it may be useful for those who are searching for a general overview of glass ceramics materials and its applications in different fields.

One of the most important themes in novel materials research is to develop innovative materials to use in solving problems in human health care. This has led to the development of synthetic materials known as biomaterials. Their fabrication for bone implants and tissue engineering, followed by their implantation in human body is a highly interdisciplinary subject to be addressed jointly by materials scientists, engineers as well as by surgeons.

The chapter one gives an introduction about the substances that can be termed biomaterials, their requirements and material-tissue interactions. Further, the development of advanced dental material technologies has recently led to the introduction of a range of all-ceramic restorations in dentistry.

In chapter two, the authors summarise the fundamental principles of glass-ceramic technology, particularly its use in dentistry, and also give general information about current commercial materials and materials currently under development. Discussing their properties, processing methods, and how they may affect the future of dentistry.

In chapter three, the microwave heating technique, which has attracted considerable attention for the processing of various materials such as ceramics, glasses, polymers, composites and even metals is reviewed. Researchers are trying to apply this technology to new areas.

Lanthanide-doped composites have been reported to have an adequate biocompatibility, further enhancing cell adhesion and proliferation, a behaviour that suggests a prospective application in bone tissue engineering approaches and these are discussed in chapter four.

Chapter five aims to address the clinical application of bone grafts on periodontal regenerative approaches, with relevance given to the use of calcium phosphate ceramics. A clinical case study is presented in which the regenerative capability of a glass-reinforced hydroxyapatite, Bonelike®, is thoroughly evaluated by clinical and tomographic measurements, in the healing of a periodontal intrabony defect.

In the final chapter, the structural and optical analysis of erbium ion doped zinc/cadmium bismuth borate/silicate glasses is reported. This is based on the results concerning the NIR emission transition $^4I_{13/2} \rightarrow {}^4I_{15/2}$ at 1.506 μm, in Er^{3+}-doped zinc/cadmium bismuth borate/ silicate glasses, which may prove to be useful in the optical communication area.

Sooraj H. Nandyala
FCUP/INESC TEC
University of Porto, Porto
Portugal

&

José D. Santos
FEUP, University of Porto, Porto
Portugal

List of Contributors

Anthony Johnson

Academic Unit of Restorative Dentistry
School of Clinical Dentistry
University of Sheffield
Claremont Crescent
Sheffield S10 2TA, UK

Ashish Agarwal

Department of Applied Physics
Guru Jambheshwar University of Science & Technology
Hisar-125001 Haryana, India

Debabrata Basu

Bio-Ceramics and Coating Division
CSIR – Central Glass and Ceramic Research
Institute, Kolkata–700 032, West Bengal, India

Pavan K. Gudi

Department of Periodontics
Govt. Dental College and Hospital
Hyderabad – 500 012 AP, India

Hawa Fathi

Academic Unit of Restorative Dentistry
School of Clinical Dentistry
University of Sheffield
Claremont Crescent
Sheffield S10 2TA, UK

Inder Pal

Department of Applied Physics
Guru Jambheshwar University of Science & Technology
Hisar-125001 Haryana, India

João Coelho

INESC Porto/ Departamento de Física
Faculdade de Ciências
Universidade do Porto
Rua do Campo Alegre, 687
4169-007 Porto, Portugal

José D. Santos

DEMM, Faculty of Engineering
University of Porto
Rua Dr. Roberto Frias
4200-465 Porto, Portugal

Kalyan S. Pal

Bio-Ceramics and Coating Division
CSIR – Central Glass and Ceramic Research
Institute, Kolkata–700 032, West Bengal, India

Mahender P. Aggarwal

Guru Nanak Khalsa College
Yamunanagar-135001, Haryana, India

Maria A. Lopes

DEMM, Faculty of Engineering
University of Porto
Rua Dr. Roberto Frias
4200-465 Porto, Portugal

Maria H. Fernandes

Laboratório de Farmacologia e Biocompatibilidade Celular
Faculdade de Medicina Dentária
Universidade do Porto
Rua Dr. Manuel Pereira da Silva
4200-393 Porto, Portugal

Mónica P. Garcia

Laboratório de Farmacologia e Biocompatibilidade Celular
Faculdade de Medicina Dentária
Universidade do Porto
Rua Dr. Manuel Pereira da Silva
4200-393 Porto, Portugal

Pannapa Sinthuprasirt

Faculty of Dentistry
Thammasat University
Kong Luang, Pathumthani, Thailand 12121

Pedro S. Gomes

Laboratório de Farmacologia e Biocompatibilidade Celular
Faculdade de Medicina Dentária
Universidade do Porto
Rua Dr. Manuel Pereira da Silva
4200-393 Porto, Portugal

Rayasa R. Rao

Materials Science Division
National Aerospace Laboratories, CSIR,
Bangalore – 560 017, India

Sarah Pollington

Academic Unit of Restorative Dentistry
School of Clinical Dentistry
University of Sheffield
Claremont Crescent
Sheffield S10 2TA, UK

Sooraj H. Nandyala

FCUP/INESC Porto/ Departamento de Física
Faculdade de Ciências

Universidade do Porto
Rua do Campo Alegre, 687
4169-007 Porto, Portugal

Someswar Datta

Bio-Ceramics and Coating Division
CSIR – Central Glass and Ceramic Research
Institute, Kolkata–700 032, West Bengal, India

Sumana Ghosh

Bio-Ceramics and Coating Division
CSIR – Central Glass and Ceramic Research
Institute, Kolkata–700 032, West Bengal, India

Sujata Sanghi

Department of Applied Physics,
Guru Jambheshwar University of Science & Technology
Hisar-125001 Haryana, India

2

Send Orders of Reprints at bspsaif@emirates.net.ae
Current Trends on Glass and Ceramic Materials, 2013, 3-48

CHAPTER 1

Current Trends in Processing and Shaping of Bioceramics

Rayasa R. Rao[*]

Materials Science Division, CSIR, National Aerospace Laboratories, Bangalore – 560 017, India

Abstract: In the recent past advanced materials with innovative processing techniques are being developed for applications starting from day to day home appliances to space shuttle. One of the most important and direct usage of advanced materials is for solving problems in human health care, particularly in replacement of the damaged or lost bone tissues. In this direction, the development of synthetic materials known as biomaterials, fabrication of their structures for bone implants and for tissue engineering (Scaffolds) followed by their implantation in human body is a highly interdisciplinary subject and is addressed jointly by material scientists, engineers as well as by the surgeons.

This chapter gives an introduction about the advanced materials that are used as biomaterial, their requirements and materials-tissue interactions. This is followed by the discussion on bioceramic materials and their classification as nearly inert, bioactive and resorbable materials with examples. Among the number of materials developed for bio-applications, those showing higher compatibility with the tissues and which proliferate the growth of tissues play a prominent role. Among these, some of the ceramics like hydroxyapatite and tricalcium phosphate are widely used due to their chemical similarity to bone and good biocompatibility. In this context, a review on preparation methods, processing and forming, thermal stability and densification and some of the characteristic properties of hydroxyapatite ceramics has been presented.

The chapter also deals with the processing and shaping aspects of bioceramic materials including basic principles, experimental result and discussions. Colloidal processing, slip casting, gel casting and mouldless casting methods are discussed as applied to Al_2O_3, hydroxyapatite and tricalcium phosphate as specific examples of bioceramic materials to fabricate differently shaped dense and porous samples intended for implant and scaffold applications.

Keywords: Bioceramics, Bioactive ceramics, Resorbable ceramics, Calcium phosphate, Hydroxyapatite, Alumina, Synthesis of HA, Thermal stability of HA, Processing of HA, Shaping of bioceramics, Nano hydroxyapatite, Colloidal processing, Zeta potential, Viscosity, Slip casting, Gel casting, Solid freeform fabrication, Porous ceramics, Scaffold, Osteoporosis, Tissue engineering.

*Address correspondence to **Rayasa R. Rao:** Materials Science Division, CSIR - National Aerospace Laboratories, P.B. No. 1779, Bangalore 560017, India; Tel: 91 8025086242; Fax: 91 8025270098; E.Mail: rrrao@nal.res.in

Sooraj H. Nandyala and José D. Santos (Eds)
All rights reserved-© 2013 Bentham Science Publishers

1. INTRODUCTION TO BIOMATERIALS

One of the most important technological advancements in the area of advanced materials is their application in improving human welfare and health. These materials generally referred to as **"biomaterials"** and are defined as "materials (non-living materials) intended to interface with human biological systems in order to repair, augment or replace any tissue or organ diseased, damaged or worn out from the body or to aid the functioning of any system which is a part of the body". Most of the known biomaterials have rarely been designed and developed specifically for this purpose. They are usually spin offs from the materials developed for other applications.

The structure and properties of biomaterials are tend to be similar to the materials developed for chemical, structural, aerospace and nuclear applications. On the other hand, the physiological environment in a human body is extremely hostile, sensitive and unforgiving when forced to accept foreign bodies. Hence, safe effective and reliable use of any synthetic material in a human body requires specific and unique properties that are found only in a few materials and are called as biomaterials.

During the ageing process, some of the problems encountered necessitate a surgical intervention and implantation of a biomaterial into the body, *e.g.* developmental defects due to impaired hormonal functioning, diseases resulting in loss of vital functions of the organs, psychological factors arising due to abnormalities of the body organs and tissue atrophy leading to change in shape and consistency of tissues. Also, congenital (birth) defects and trauma through accidents requires surgical intervention. Several synthetic biological materials or biomaterials are developed to meet major clinical requirements which broadly fall under (1) Orthopaedic surgery concerned with the musculoskeletal systems involving treatment of bones, cartilage, ligaments, muscles tendons and joints (hip and knee joints) (2) Maxillofacial surgery concerned with the structural tissues of the head involving bone and soft tissue, *e.g.* dental implants, internal fracture plates for mandible or maxillary bones *etc.* (3) Cardiovascular surgery connected with heart and blood vessels. (4) Ophthalmology which makes use of artificial lens, artificial cornea, contact lenses, artificial eye ball *etc.* (5) Nervous system mainly concerned with the regeneration and reconstitution of damaged nerve tissue [1-4].

1.1. Requirements for Biomaterials

The biomaterials used for various applications have to perform under variety of functional and biological environments. Generally, the requirements for a biomaterial can be divided into those properties, which determine suitability of a material to perform specific functions and those, which determine its acceptability within the body. Thus a biomaterial must have both biomechanical and biochemical compatibility with the body tissues.

With respect to the mechanical performance, the biomaterial required for soft tissue reconstruction is different from the materials that are needed for hard tissues and the yield and fracture strengths of the material-tissue combination are crucial. Ideally, the rigidity, flexibility, compliance or any other elastic deformation properties should be similar to the replaced natural component. Due to the chemically hostile environment of the body, mechanochemical behaviour such as stress-corrosion, cracking, corrosion-fatigue, composite-delamination, stress-crazing *etc.* becomes important. Besides this, other physical properties, such as density, porosity, optical and electrical properties need to be addressed.

The long term performance of the biomaterials significantly depends upon the chemical and biochemical interactions between biomaterials and their host. These interactions termed, as "biocompatibility" is referred as the ability of a material to perform with an appropriate host-response in a specific application. This phenomenon includes a sequence of reactions and effects that take place between the material and the system (body) [1, 4-8].

1.2. Material - Tissue Response – Types of Biomaterials

All materials implanted in the living body elicit a response from the host tissue at the tissue - implant interface. This response is influenced by various factors like, type, health and age of tissue, blood circulation in tissue and at the interface *etc.* from the tissue side and on the other hand composition, phases and phase boundaries, surface morphology, surface porosity, chemical reactions, movements at interface, closeness of fit, mechanical load *etc.* from the implant side. The mechanism of tissue attachment to an implant is related to the characteristics of the tissue and the implant as well as to the implant - tissue interaction occurring at the interface. The primary requirement for an implant material is that it should not have any toxic effect. Toxicity kills the surrounding cells or produces systemic damage to the patient. Ceramic implants are of interest in this regard because of their inertness and lack of toxicity [6].

A **Type I biomaterial** is biologically nearly inert material which does not form any chemical or biological bond with the tissues. Such a material is attached by bone growth into surface irregularities by cementing the device into the tissues or by press fitting into a defect, termed as "morphological fixation". The most common tissue response to a biologically inactive, nearly inert implant is the formation of a non-adherent fibrous capsule around the implant, which isolates the implant from the host. The thickness of the fibrous layer depends on the physico-chemical conditions existing at the implant tissue interface. As the interface is not chemically bonded, relative movement (micro-motion) can occur at the interface, which may result in deterioration of the function of the implant or of the host tissue or both. Metallic materials like titanium, stainless steel, ceramics like alumina and zirconia, polymeric materials like polyethylene are the examples of nearly inert biomaterial.

The **Type 2 porous biomaterials** allow the in-growth of tissue into pores present on the surface or on the interior of the implant. This type of attachment is called biological fixation and it provides a mechanical bonding with large interfacial area between the implant and the host. Thus, it is capable of withstanding more complex stress than type 1 implants. The limitation is that the pores must be greater than 100 µm in diameter so that capillaries can provide a blood supply to the ingrown connective tissues. Porous biomaterials may be made of metallic, ceramic as well as polymeric materials or their combinations as the case may be.

The **Type 3 bioactive material** elicits a specific biological response which involves a series of biophysical and biochemical reactions at the implant tissue interface. This leads to the formation of a strong chemical interfacial bond between the tissues and the material called as bioactive fixation. Thus, a bioactive material creates an environment compatible for osteogenesis (bone growth). The mineralizing interface develops a natural bonding junction between the living body and the non-living biomaterial implant. This phenomenon is referred as "bioactivity". This interfacial bond prevents motion between the two materials and mimics the type of interface that is formed when natural tissues repair themselves. Further, the bioactive interface is in a state of dynamic equilibrium and change with time, as do natural tissues.

The mechanism and the time dependence of bonding, the strength of the bond, the thickness of bonding zone *etc.*, varies for different materials and the rate of development of the interfacial bond can be referred as the level of bioactivity. A common characteristic of all bioactive implants is the formation of a biologically active hydroxy-carbonate apatite (HCA) layer on their surface, which is equivalent in composition and structure to the mineral phase of the bone. Synthetic hydroxyapatite, bioactive glass and glass ceramics are well known bioactive materials for a variety of biomedical applications.

The **Type 4 resorbable materials** degrade or dissolve gradually with time and replaced by natural host tissues. The degradation products must be chemical compounds that are not toxic and can be easily disposed by the system without any damage to the cells (metabolically acceptable materials) and the resorption rate must match to the repair rates of body tissues. This leads to the regeneration of tissues instead of their replacement. The unique advantage is that initial pore size of the implant can be small thereby possessing high mechanical strength compared to that of more porous substances. As the material dissolves, porosity increases and thus allows the in-growth of supporting tissue. This in-growth gives the mechanical integrity to the implant. The requirement of short-term mechanical performance, the matching of resorption rates to the repair rates and metabolically acceptable constituents are some of the severe demands on this type of implant material. Tricalcium phosphate and polylactic acid are promising examples in this category.

In principle, no single biomaterial is entirely optimal for all applications. It is necessary to match the form, type and properties of a material with its rate of bonding and other functions in the body. Interestingly, small changes in composition can make a biomaterial nearly inert, resorbable or bioactive.

Based on the material categories, the biomaterials include metals and alloys, polymers, ceramics and glasses, carbons and composites. The detailed knowledge on these materials can be obtained from various review articles [1-8]. Because of the significance and relevance in the present context, the discussion is restricted to ceramic materials that are extensively used for bio-applications.

1.3. New Generation Biomaterials

The development of first generation (bioinert) and second generation, (bioactive and resorbable) biomaterials followed by their clinical success benefited millions of patients. However, most of the skeletal prosthesis failed within 10 to 25 years and patients required a revision in surgery. This is due to the fact that synthetic materials are not able to respond to change in physiological loads or biochemical stimuli, unlike living tissues and this, limits the lifetime of artificial body parts. Emphasis on the development of more biologically based method for the repair and regeneration of tissues resulted in the development of third generation biomaterials.

Third generation biomaterials are being designed to stimulate specific cellular responses at the level of molecular biology. In this approach, two alternative routes are followed. (1) In tissue engineering progenitor cells are seeded on scaffolds outside the body, where they grow, differentiate and mimic naturally occurring tissues. These tissue engineered constructs are then implanted into patients wherein the scaffolds are resorbed and replaced by host tissues with a viable blood supply and nerves with time. The living tissue engineered constructs adapt to the physiological environment to provide a long lasting repair. (2) In tissue regeneration approach, bioactive materials are used as powders, solution or doped microparticles release chemicals as ionic dissolution products at controlled rates. These released chemicals act as macromolecular growth factors and activate the cells in contact to produce additional growth factors. These stimulations lead to multiple generations of growing cells and their self assembly into the required tissues. Bioactive glass, composites, macroporous scaffolds, gel-glass foams are the some of third generation biomaterials [9].

2. BIOCERAMICS

The clay is irreversibly transformed by fire into ceramic pottery which was discovered long back led to the transformation of human culture from nomadic hunters to agrarian settlers. From those days, ceramic pots were used to store grains for a long period with minimal deterioration and impervious ceramic vessels were used for cooking. During the last five decades, another revolution has occurred in the use of ceramics to improve the quality of life with the development of specially designed and fabricated ceramic materials for biomedical applications. Ceramics and glasses have been used for a long time outside the body for a variety of health care applications. Ceramics are widely used in dentistry as restorative materials, gold-porcelain crowns, glass-filled ionomer cements, in endodontic treatments, dentures *etc.* In recent years different ceramic materials have found applications as bone replacement materials, in tissue engineering, as prosthesis as well as in drug delivery systems. Bioceramics are produced in a variety of forms and phases depending upon the desired properties and required functions for repair and reconstruction of specific parts of the body. The detailed information on various types of bioceramic materials with respect to chemical composition, form, phase, functions and clinical use is given elsewhere [10-18].

2.1. Nearly Inert Crystalline Bioceramics

Alumina (Al_2O_3) and zirconia (ZrO_2) are the widely used inert ceramics for biomedical applications. High density and high purity (> 99.5%) α-Al_2O_3 was the first clinically used bioceramic. It is used in load-bearing hip prosthesis (heads of artificial hip) and in dental implants because of its excellent properties such as good corrosion resistance, good biocompatibility, high wear resistance and high strength. The main advantages of Al_2O_3 in hip prosthesis are its excellent wear rate characteristics and its low coefficient of friction, the factors which maximize the functional capabilities of artificial hip. The limitation in the use of alumina is the stress shielding of surrounding bone due to its significantly higher elastic modulus (380 GPa) as compared to that of cortical bone (72.5 GPa). Medical grade alumina developed in accordance with ISO 6474 is used for the production of weight bearing components for hip endoprosthesis. Other clinical applications of alumina include, knee prosthesis, bone screws, jaw bone reconstruction, middle ear bone substitutes, corneal replacements, segmental bone replacements and blade, screw or post type dental implants etc [4, 19].

ZrO_2 offers very attractive mechanical properties, specially, high fracture toughness and biocompatible as alumina. Femoral heads made of zirconia are introduced about two decades ago in France, USA and Australia. Magnesia partially stabilised zirconia (Mg-PSZ) or Yttria stabilised tetragonal zirconia polycrystals (Y-TZP) are used for medical applications and Y-TZP ceramics has ISO 13356 standards. Titania (TiO_2), silica (SiO_2), calcium aluminates (CaO-Al_2O_3) and aluminosilicates (Al_2O_3-Na_2O-SiO_2) are other inert oxide ceramics for bio-applications. While the non-oxide material like carbon (in vitreous or pyrolytic form) and graphite are widely used as biomaterials, the nitride and carbide ceramics (Si_3N_4, SiC) are also used for biomedical applications [4, 20].

2.2. Bioactive Ceramics

Hydroxyapatite {HA; $Ca_{10}(PO_4)_6(OH)_2$} belongs to a group of calcium phosphate ceramics that are used for more than 25 years as bone substitute materials [21, 22]. HA is chemically stable inside the tissues and is bioactive. While HA has the advantage of excellent biocompatibility, its inferior mechanical properties leads to a limitation with respect to the functional reliability, specially, under tensile load. Therefore, these can only be used as powders for space filling or as small non-load bearing implants as in the middle ear, as dental implants with reinforcing metal posts, as coatings on metal implants. HA is also used as low-load bearing porous implants where interpenetrating bone growth acts as reinforcing phase. Nano particles of HA are in use for controlled drug delivery applications. The 3-dimensional structures having multidimensional porosity made of HA, HA-β-TCP biphasic ceramics and HA-polymer systems are useful as scaffolds in tissue engineering applications.

Bioactive glass, the first bioactive material that has the capability of bonding to bone was developed by Hench and coworkers in 1972 [7, 8]. This contained SiO_2, Na_2O, CaO and P_2O_5 in specific proportions, and called as **bioglass.** A series of bioglasses are developed with four component systems that have compositional features of SiO_2 < 60 mol %, high Na_2O (20-25%) and high CaO (20-25%) content, high CaO/P_2O_5 ratio and a constant 6 wt% P_2O_5. The low silica content and the presence of calcium and phosphate ions result in very rapid ion exchange in physiological solutions and a rapid nucleation and crystallisation of hydroxyl carbonate apatite bone mineral on the surface. This growing bone mineral layer bonds to collagen fibrils produced by the bone cells forming a strong interfacial bond between the implant and the living tissues. Different bioglasses have different rates of bonding which depends on their composition. The low mechanical property restricts its clinical application to non load bearing situations such as middle ear implants, alveolar ridge reconstruction and particulates for periodontal, maxillofacial and orthopedic repair. Bioglass can be used as coating on tougher materials (metal alloys or ceramics) for dental implants and hip prosthesis.

Bioactive glass - ceramics are developed by the process of controlled crystallisation of glass which displays bone bonding phenomena similar to bioglass but with improved mechanical properties. A low alkali (0-5wt%) bioactive silica glass ceramics known as Ceravital and a two phase silica-phosphate glass-ceramic composed of 20% HA and 55% wollastonite crystals in a residual SiO_2 glassy matrix (A/W glass-ceramic) are specific examples. An addition of Al_2O_3, TiO_2, Ta_2O_5, Sb_2O_3 or ZrO_2 is found to inhibit bone bonding while the second phosphate phase like β-whitlockite [$Ca_3(PO_4)_2$] does not. Another variation is the machinable glass-ceramic consisting of mica (provide machinability) and apatite (provide bioactivity) crystals within the glassy matrix.

2.3. Resorbable Bioceramics

The function of a resorbable bioceramic is of transient nature. It provides a temporary space filling scaffold until natural tissues grow gradually and totally replace the implant. Resorbable bioceramics are used to treat maxillofacial defects, for obliterating periodontal pockets, as artificial tendons and as composite bone plates. Plaster of Paris (Calcium sulfate) was one of the first materials evaluated as a resorbable bone substitute. Its usage became limited because of its lower strength and unpredictable rate of resorption. Trisodium phosphate is another biocompatible resorbable ceramic material. Among various calcium phosphate ceramics, β-tricalcium phosphate (β-TCP) is highly biodegradable while hydroxyapatite shows good stability under physiological conditions. When implanted, TCP will interact with body fluids to form HA as per the following reaction (1):

$$4Ca_3(PO_4)_2 \text{ (s)} + 2H_2O \rightarrow Ca_{10}(PO_4)_6(OH)_2 \text{ (surface)} + 2Ca^{2+} + 2HPO_4^{2-} \qquad \textbf{(1)}$$

Thus, the solubility of TCP surface approaches the solubility of HA and the pH of the solution decreases further, which increases the solubility of TCP and enhance resorption. The solubility of calcium phosphates is a function of Ca/P ratio. The microporosities in the sintered material can increase the solubility of these phases. Theoretically, resorbable TCP is an ideal implant material. However, the lower mechanical performance and metabolic problems that arises by the release of large size grains are the main limitations. The tissue compatibility and repair process with tricalcium phosphate ceramic is superior to other synthetic materials and is equivalent to or even better than autogenous bone for repairing marginal periodontal defects [23-30].

2.4. Bioceramic Composites and Coatings

While the bioceramic materials have the advantage of good biocompatibility, most of the bioactive ceramics have low biomechanical compatibility than required for load bearing applications. They have uncertain lifetimes under the complex stress states, slow crack growth and cyclic fatigue that arise in many clinical applications. The principle reason behind this is the compositional and structural difference between the biological tissues (bone, enamel and dentine) and the synthetic implant materials, which replaces them. Structurally bone is a composite of an organic substance (primarily collagen) and an inorganic substance (primarily biological apatite; HCA). Hence, the structural tailoring of bioceramic materials has been approached through the development of biocomposite materials or through biomaterial coatings [4, 7, 10, 11, 31-33].

The bioceramic composite involves all the three types of biomaterials - nearly inert, bioactive and resorbable. They are composed of polymer, carbon, glass or ceramic matrix reinforced with various types of fibres or particulates including carbon, SiC, stainless steel, HA, phosphate glass, Al_2O_3, alumino silicates and ZrO_2. The purpose is to increase the flexural strength, strain to failure (fracture toughness) and to decrease the elastic modulus. This can be achieved by two approaches. The first approach is to reinforce the ceramic, glass or glass-ceramic matrix with high fracture toughness phases like metal fibres or with ceramic particles and ceramic fibres. Bioglass reinforced with stainless steel fibres (60Vol%), HA reinforced with particles or fibers of ZrO_2 (Y_2O_3), apatite-wollastonite glass ceramic containing dispersion of tetragonal zirconia are some examples of composites having higher strength and toughness. However, these composites have high elastic modulus than the bone and thus give rise to stress shielding of the bone. Another approach pioneered by Bonefield *et al.* [11] is to use bioactive glass, or ceramic particles or fibers to reinforce an elastically compliant and biocompatible polymer matrix. The matrix includes polymers such as polyethylene, PMMA, polysulfone or collagen and the reinforcing phases are HA, calcium phosphates and bioactive glasses as powders or fibers. This polymer - bioactive ceramic composite solves the problem of stress shielding since its young's modulus is ideally matched to that of the bone.

An alternative approach of achieving biomechanical compatibility of ceramics for load-bearing applications is coating of a bioceramic material on to a mechanically tough substrate. Metals and medical grade alumina either in the dense or porous form are the commonly used substrates. HA, TCP, various bioglass compositions and pyrolytic carbon are the commonly used coating materials. Various techniques are available for achieving the coating including hot isostatic pressing, plasma or flame spraying, ion-beam sputtering, magnetron sputtering, frit enamelling, electrophoretic deposition, sol-gel deposition and rapid immersion coating. The bone-bonding capacity of these coatings provides cementless fixation of orthopedic prosthesis and enhances the early stage interfacial bond strength. However, they have deficiencies with respect to reliability of the coating/implant interface for long-term implantations.

3. HYDROXYAPATITE CERAMICS

A calcium phosphate reagent described as 'triple calcium phosphate compound' was successfully used for the repair of bone defects as early as 1920. About half a century later, methods have been reported for synthesizing calcium phosphate ceramics from mineral fluorapatite $\{Ca_{10}(PO_4)_6F_2\}$ or other commercially available calcium phosphate reagents and using them for clinical applications. These ceramics were found later to be a mixture of β-TCP and HA. In the mid seventies, three groups of scientists, Jarcho *et al.* [34] in U.S.A, de Groot *et al.* and Denissen [35] in Europe and Aoki *et al.* [36] in Japan worked independently on the development and commercialisation of HA as a biomaterial for repair, augmentation or substitition of the bones and skeletal systems of the body. Their pioneering research has drawn the attention of researchers world over and as a result, HA in the form of granules were approved for medical use by the German Federal Board of Health in May 1990 and American standard for medical grade HA namely ASTM F 1185 were drawn up in 1988 [19-21], edited in 2003 as F1185-03 and re-approved in 2009. It stipulates that the mineral phase must contain minimum 95% of HA and maximum allowable heavy metal and other impurities limited to only 50ppm. A number of state-of-the-art articles are available concerning the preparative methods, thermal stability, sintering and the characteristic properties of HA and other calcium phosphate ceramics and on their uses as biomaterials [4, 8, 12, 22-30, 37]. In the following sections, a brief review of some of the physicochemical aspects of HA are discussed.

3.1. Methods for Synthesizing HA

HA in the pure form or in combination with other calcium phosphates can be synthesized by; (i) Wet methods involving precipitation from aqueous solutions containing calcium and phosphate ions or by hydrolysis of acid calcium phosphates, (ii) Dry methods involving high temperature solid state reactions between calcium and phosphatic compounds, (iii) Hydrothermal methods involving the simultaneous application of high temperature and high pressure to aqueous solutions of calcium and phosphate to precipitate HA and (iv) Miscellaneous methods such as sol - gel, polymeric precursor, solution combustion, mechanochemical, microwave assisted, sonochemical *etc.*

The precipitation of calcium phosphates from aqueous solutions of calcium and phosphate ions is significant since various calcium phosphates are present as mineral constituents of calcified tissues like bone and teeth. Also, the precipitated product having compositional variations become the precursor for other preparative methods.

3.1.1. The Phase Diagram CaO - P₂O₅ - H₂O

The conditions for the formation of calcium phosphates and their stability can be established in terms of the phase diagram between the constituent oxides CaO, P_2O_5 and H_2O. The ternary phase diagram of the CaO - P_2O_5 - H_2O system is depicted in Fig. **1** [37, 38].

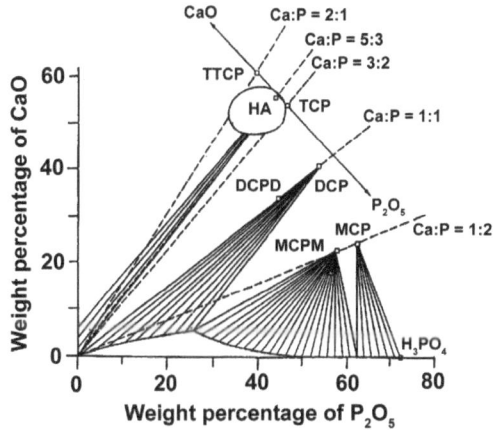

Figure 1: Ternary phase diagram of the CaO - P_2O_5 - H_2O system constructed at 25°C [37, 38]; MCP = Anhydrous monocalcium phosphate [$Ca(H_2PO_4)_2$], MCPM = Monocalcium phosphate monohyadrate [$Ca(H_2PO_4)_2.H_2O$], DCP = Anhydrous dicalcium phosphate [$CaHPO_4$]; DCPD = Dicalcium phosphate dihydrate [$CaHPO_4. 2H_2O$]; TCP = Tricalcium phosphate [$Ca_3(PO_4)_2$]; TTCP = Tetracalcium phosphate [$Ca_4P_2O_9$], HA = Hydroxyapatite [$Ca_{10}(PO_4)_6(OH)_2$].

Anhydrous monocalcium phosphate [$Ca(H_2PO_4)_2$ (MCP)], its monohydrate [$Ca(H_2PO_4)_2.H_2O$ (MCPM)] (Ca:P = 1:2) and anhydrous dicalcium phosphate [$CaHPO_4$ (DCP)] (Ca:P = 1:1) exist as stable phases in the acidic conditions. Dicalcium phosphate dihydrate [$CaHPO_4. 2H_2O$ (DCPD)] exists as a metastable phase over a limited range of experimental conditions. The hydroxyapatite (HA) phase, a crystalline precipitate showing the X-ray diffraction pattern similar to that of an apatite with the general formula $Ca_{10} (PO_4)_6 (OH)_2$ is characterised by a variable composition with Ca to P ratio ranging from 1.5 to 2 (Ca:P = 3:2 to 2:1). The limiting ratios of 1.5 and 2 corresponds to tricalcium phosphate {TCP; $Ca_3(PO_4)_2$} and tetracalcium phosphate {TTCP; $Ca_4P_2O_9$}, respectively. Thus, HA can be viewed as a defect structure which exists over a compositional range Ca_{10-x} $(HPO_4)_x (PO_4)_{6-x} (OH)_{2-x}$ where x ≤1. The non-stoichiometry is accomplished partly by the introduction of vacancies. The composition of calcium deficient or non-stoichiometric HA (CDHA or ns-HA), $Ca_9 (HPO_4)(PO_4)_5$ OH having Ca/P ratio of 1.5 is achieved by the removal of CaO from the HA structure (x=1) resulting in calcium and oxygen vacancies. The CDHA is more acidic than stoichiometric HA (s-HA) and coexist in equilibrium with $CaHPO_4$, while it forms metastable invariant point with $CaHPO_4.2H_2O$ (the phases more acidic than HA). The formation of s-HA (x=0) is favored at high pH and it forms an invariant point with $Ca(OH)_2$. HA is the only solid phase capable of existence in the alkaline range and the phases stable in the acid region can be transformed into HA through hydrolysis. An additional metastable solid phase called octacalcium phosphate {OCP, $Ca_8H_2(PO_4)_6 \, 5H_2O$} can exist in the in the neutral region, which could be identified by X-ray diffraction has not been indicated in the above phase diagram. The calcium phosphate phases like calcium pyrophosphate (CPP), tricalcium phosphate (TCP), and tetracalcium phosphate (TTCP) which are formed at high temperatures are considered while discussing the thermal stability/densification of HA.

3.1.2. Wet Methods

Precipitation Method: Precipitation from aqueous solutions is the most commonly used method for commercial preparations of HA and TCP. A method developed by Hayek *et al.* [39] and later modified by Jarcho *et al.* [34] is based on the reaction between calcium nitrate $Ca(NO_3)_2$ and diammonium hydrogen phosphate $(NH_4)_2HPO_4$ with pH adjusted in the range 11-12 by addition of NH_4OH solution according to the reaction (2);

$$10Ca(NO_3)_2 + 6 \, (NH_4)_2HPO_4 + 8NH_4OH \rightarrow Ca_{10}(PO_4)_6(OH)_2 + 20NH_4NO_3 + 6H_2O \qquad \textbf{(2)}$$

The accompanying volatile ammonium salts are sublimed off by heating the filtered product to 250°C. Jarcho *et al.* [34] investigated the effect of initial Ca/P ratio and stirring time on the relative amounts of HA and β-TCP formed in the sintered samples. When producing 100% pure β-TCP, the catalytic effect of SO_4^{2-} has been extensively studied. Since then, this method is investigated in depth by many groups [40-42]. The effect of stirring time, initial mixing Ca/P ratio, temperature, reactant concentrations, rate of addition, residence time and pH of solutions on the nucleation and growth kinetics, particle size distribution of the precipitates, phase composition and Ca/P ratio of the final products as well as their sintering behaviour and mechanical properties are all well studied.

Many modifications of the precipitation method have been carried out with minor alterations to produce isomorphs of HA. Ca could be substituted by Ba, Sr, Mg, Mn, Cd, Pb, Zn *etc.*, while the phosphate can be replaced with carbonate, vanadate, borate, manganate, arsenate *etc.*, and OH⁻ ions by fluoride, chloride and carbonate. Use of calcium acetate as the calcium source and sodium or potassium hydrogen phosphates as the phosphate source have also been investigated by many researchers [4, 22, 43-47].

Another precipitation method consists of drop wise addition of phosphoric acid to a continuously stirred suspension of $Ca(OH)_2$ in water and maintaining the alkaline pH (pH>7) by the addition of NH_4OH solution as per the reaction (3);

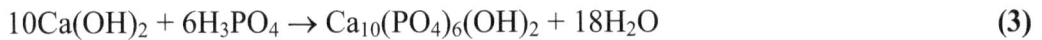

$$10Ca(OH)_2 + 6H_3PO_4 \rightarrow Ca_{10}(PO_4)_6(OH)_2 + 18H_2O \qquad (3)$$

This method is more suitable for industrial production of HA since the only by-product is water. Osaka *et al.* [48] examined the effect of the reaction temperature, the pH of the system, rate of mixing of solutions, and mixing Ca/P molar ratio, on the stoichiometry and properties of the precipitated product. They observed that s-HA of very fine particles suitable for dense ceramics is obtained by slower addition of the phosphoric acid at lower temperatures followed by using appropriate ripening procedures. The ns-HA can be readily precipitated because the pH is lowered in the vicinity of HA particles where HPO_4^{2-} ions are more stable than PO_4^{3-} ions. Similarly, Slosarczyk *et al.* [49] examined the relationship between the Ca:P molar ratio of the initial precipitates and the phase composition as well as the properties of the sintered products such as density, sintering shrinkage, hardness, bending strength and the roughness of their fractured surfaces. This method of HA synthesis has also been employed for obtaining HA doped with various ions.

In contrast to the above precipitation methods, where the composition of the mixing reagents change with respect to time, a precipitation method under conditions of constant solution concentrations for long periods of time has been developed and studied for the growth of both s-HA and ns-HA. These conditions approach more closely those of real biological systems [50]. Brown's group discovered the formation of CDHA and s-HA from acid - base reaction between DCP or MCPM and TTCP at near physiological temperatures [51, 52].

Hydrolysis Method: HA can also be prepared by the hydrolysis of acidic calcium phosphates like DCPD (brushite, $CaHPO_4. 2H_2O$), OCP $\{Ca_8H_2(PO_4)_6 5H_2O\}$ and DCP (monetite, $CaHPO_4$) in hydroxide, carbonate, fluoride or chloride salt solutions of ammonium, sodium or potassium depending on the composition of the HA desired. Calcium carbonate can also be hydrolysed to HA in ammonium or sodium phosphate solutions or to fluorapatite in fluoride solutions. α- or β-TCP, TTCP and amorphous calcium phosphate (ACP) can also be easily hydrolysed to CDHA [22, 53].

Generally, the HA prepared from aqueous systems by precipitation or hydrolysis are calcium deficient and HPO_4^- enriched resulting in the formation of β-TCP on heat treating at $\geq 800°C$. If the reaction is carried out under very basic conditions, the Ca/P ratio approaches that of stoichiometric value (1.67) or exceeds due to the formation of carbonate apatite by substitution of carbonate ion from solution. The commercially available calcium phosphate reagents labelled or mislabelled as "Calcium phosphate tribasic" are some times mixed phases of apatitic calcium phosphate and monetite [54, 55] and reagents labelled as "spheroidal hydroxyapatite" consist mostly β-TCP mixed with small amounts of HA.

3.1.3. Dry Method

In this method, HA is synthesised by a solid state reaction between appropriate solid ingredients at optimum high temperatures. HA is formed from DCP, TCP, TTCP and $Ca(OH)_2/CaO/CaCO_3$ taken in appropriate proportions to give Ca to P molar ratio of $5/3 = 1.67$ corresponding to s-HA [22, 37] as per the following reactions (4-7):

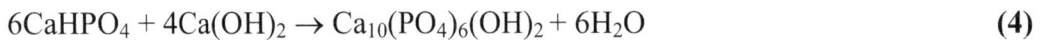

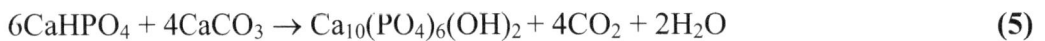

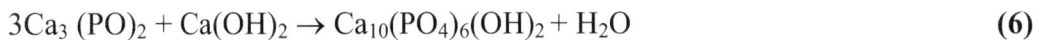

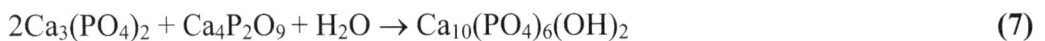

$$6CaHPO_4 + 4Ca(OH)_2 \rightarrow Ca_{10}(PO_4)_6(OH)_2 + 6H_2O \tag{4}$$

$$6CaHPO_4 + 4CaCO_3 \rightarrow Ca_{10}(PO_4)_6(OH)_2 + 4CO_2 + 2H_2O \tag{5}$$

$$3Ca_3(PO)_2 + Ca(OH)_2 \rightarrow Ca_{10}(PO_4)_6(OH)_2 + H_2O \tag{6}$$

$$2Ca_3(PO_4)_2 + Ca_4P_2O_9 + H_2O \rightarrow Ca_{10}(PO_4)_6(OH)_2 \tag{7}$$

This method has been extended to prepare chlor-apatite and its series of solid solutions with HA as well as to prepare various isomorphs of HA with substitutions of As, Ba *etc.* [37]. Nordstrom *et al.* [54] prepared an apatite slurry by mixing stoichiometric amounts of TCP and $Ca(OH)_2$ in water which after optimisation was cast into plaster moulds. The green bodies were heated in the range 1050 to 1300°C to produce sintered HA. The so formed HA was further doped with carbonate ions by soaking the crushed HA in ordinary mineral water for different time intervals. Ramachandra Rao *et al.* [55] synthesised pure HA and HA-β-TCP biphasic ceramics by solid state reaction between commercially available TCP and $Ca(OH)_2$ and studied their thermal stability in the temperature range 1000-1250°C. The high temperature solid state reaction method has been adopted for the synthesis of α- and β-TCP, TTCP as well as doped HA starting from various calcium phosphate precursors. The dry method has the advantage of producing s-HA (in comparison to wet methods) or biphasic calcium phosphate (BCP) with desired HA/β-TCP ratio by using controlled amounts of precursors [22, 55].

3.1.4. Hydrothermal Method

This method involves heating the precursors in water or moisture environment under pressurized condition in a closed vessel. Hydrothermal methods have the main advantage in achieving considerable high crystallinity and high purity of the product. HA has been synthesised by hydrolysis of DCPD and DCP (8, 9) in the temperature range of 275-375°C and pressure of ~1250psi [22, 37]. HA can also be prepared hydrothermally from various other calcium phosphates like calcium pyrophosphate (CPP; $Ca_2P_2O_7$), (10) β-TCP and TTCP *etc.*

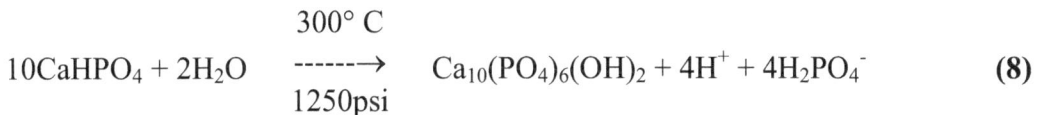

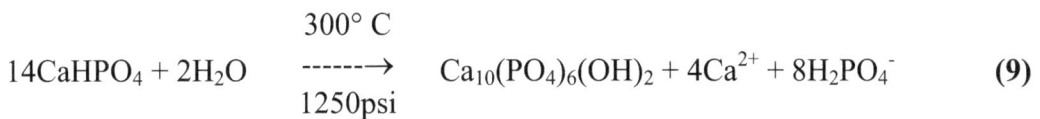

$$10CaHPO_4 + 2H_2O \quad \xrightarrow[1250psi]{300°\ C} \quad Ca_{10}(PO_4)_6(OH)_2 + 4H^+ + 4H_2PO_4^- \qquad (8)$$

$$14CaHPO_4 + 2H_2O \quad \xrightarrow[1250psi]{300°\ C} \quad Ca_{10}(PO_4)_6(OH)_2 + 4Ca^{2+} + 8H_2PO_4^- \qquad (9)$$

$$3Ca_2P_2O_7 + 4CaO + H_2O \longrightarrow Ca_{10}(PO_4)_6(OH)_2 \qquad (10)$$

By this method, HA and β-TCP were also synthesised using calcium acetate and triethyl phosphate, skeletal calcium carbonate and diammonium hydrogen phosphate as precursors of Ca and PO_4 ions, respectively. Various modifications have been employed to synthesise either powders, whiskers or platelets of HA with various morphological characteristics [56-60].

3.1.5. Miscellaneous Methods

Among the special techniques employed for the synthesis of HA, the sol-gel route has been adopted by many investigators [61, 62]. The other methods for HA synthesis include polymeric route [63], solution combustion method and mechanochemical synthesis [64]. Electrocrystallisation of HA from dilute electrolytes containing calcium and phosphorous ions [65, 66] provide an effective means for fabricating bioactive calcium phosphate coatings on porous or non porous substrates at low temperatures. The synthesis of HA from aqueous solutions using microwave irradiation [67] and ion beam assisted deposition of HA coatings on titanium alloy [68] are other advanced techniques.

3.2. Processing and Forming of HA

The HA powders prepared through any of the methods described above can be processed into different forms like dense or macroporous bodies, particulates or granulates, as coatings or in slurry form depending upon the applications. The dense HA, generally has a microporosity of less than 5% by volume and maximum pore size of less than 1 μm in diameter. The microporosity, unintentionally introduced is dependent on purity and morphology of the powder and the temperature and duration of sintering. On the other hand, macroporosities are deliberately introduced by mixing the powder with a volatile component during forming and then burning them off at relatively low temperatures before sintering. The pore sizes or spaces between particles must exceed 100 μm for bone in-growth to occur. Particulates in irregular or spherical shapes are obtained by milling and rolling the compacted and sintered blocks. Granulates are prepared from the powders by making use of proper binders followed by spray drying and/or calcination at high temperatures. Coating of HA is given on metallic or tough ceramic (Al_2O_3) substrates by thermal spraying, sol-gel coating, polymeric coating *etc.* [5, 19, 22].

Conventionally, the forming of a solid body involves the cold compaction (uniaxial die pressing) of HA powders mixed with binders (1-2 wt.%). The binder from the compact is removed by heating at 200-300°C and densification is achieved by sintering at the desired temperatures [7, 19, 22, 34-36, 41, 69]. The colloidal processing route has been applied through the study of the dispersion behaviour of HA powder in a liquid medium followed by slip casting to produce shaped products [54, 70-73]. The dispersion of HA microcrystals in distilled water or in an equivalent physiological salt solution has been used to give a homogeneous coating of HA on pure titanium rods by a dipping method [74].

The porous structure influences the resorption rate of the implant as well as the in-growth of bone depending upon the pore size, pore distribution and its architecture. Hence, different methods have been employed for producing porous HA ceramics. Porous HA has been obtained through uniaxial pressing as well as by a slip casting route by using various pore formers like polyvinyl butyral, polyurethane and cellulose sponges [75, 76]. Porous HA can also be obtained by using gas forming agents as well as by the hydrothermal exchange reaction between skeletal $CaCO_3$ from coral and diammonium hydrogen phosphate [77-79].

Besides the traditional uniaxial pressing and slip casting through colloidal processing, other specialised processing techniques like extrusion, cold isostatic pressing, gel casting, injection moulding, hot pressing, HIPping and solid freeform fabrication (SFF) methods have also been adopted for the production of HA ceramics [80-82]. Takagi *et al.* [83] used the combination of filtercake forming process and HIPping to prepare tetragonal zirconia polycrystal (TZP) dispersed HA composite having high strength and toughness. A visco-plastic processing method applied to HA by Shaw *et al.* [84] enhanced the sintering behaviour and show improved mechanical properties as compared to those samples produced by conventional cold pressing.

3.3. Thermal Stability and Densification of HA

From the phase diagram of CaO-P_2O_5-H_2O system [38] at room temperature reproduced in Fig. **1**, HA is seen as a compound with a variable composition represented by the general formula $Ca_{10-x}(HPO_4)_x(PO_4)_{6-x}(OH)_{2-x}$ where $x \leq 1$. Also, it has been found that the ns-CDHA is very easily precipitated due to improper precipitation conditions like low initial Ca/P ratio, high addition rates, mixing time, temperature *etc.* [22, 40, 41, 48, 49, 55-58]. However, this ns-CDHA (defective HA) and s-HA have similar crystal structures and belong to the same space group (P6$_{3/m}$) possessing identical XRD spectra. But the incorporation of a HPO$_4^{2-}$ group for PO$_4^{3-}$ group in the lattice is identified by the presence of a specific IR absorption peak at 875 cm^{-1} in the infrared spectra of the HA compound [55, 85-86].

The stoichiometry of the starting HA affect greatly on the stability of the HA phase at high temperatures and results in decomposition into various calcium phosphates. The knowledge of the stability of HA phase and decomposition products are significant to understand its sintering behavior in the temperature range 1000 to 1500°C. Further, various calcium phosphates formed at higher temperatures will show variation in mechanical, physical, physicochemical (solubility) properties and tissue interactions to a great extent when used as implant materials. Along with stoichiometry, factors such as purity, particle size and distribution of the powder, sintering temperature and conditions *etc.* are to be considered. Various studies have been undertaken to understand the high temperature stability and sinterability of s-HA and ns-CDHA with Ca/P ratios ranging from 1.5 to >1.67 [22, 41, 49, 55, 56, 86].

All the ns-CDHA with Ca/P ratio 1.5 to 1.66 are stable up to about 600 to 700°C displaying a cyrstallographic structure similar to that of s-HA. When heated above 600 to 700°C, the ns-CDHA decomposes into β-TCP and H$_2$O through two successive reactions namely, the condensation of the HPO$_4^{2-}$ ions to P$_2$O$_7^{4-}$ and H$_2$O followed by the reaction of P$_2$O$_7^{4-}$ with OH$^-$ to give PO$_4^{3-}$ and H$_2$O [55] as per the following reactions (11-13);

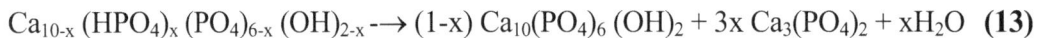

$$2HPO_4^{2-} \text{----------} \rightarrow P_2O_7^{4-} + H_2O \tag{11}$$

$$P_2O_7^{4-} + 2OH^- \text{--------} \rightarrow 2PO_4^{3-} + H_2O \tag{12}$$

$$Ca_{10-x}(HPO_4)_x(PO_4)_{6-x}(OH)_{2-x} \text{-} \rightarrow (1-x) Ca_{10}(PO_4)_6(OH)_2 + 3x\, Ca_3(PO_4)_2 + xH_2O \tag{13}$$

Reaction 13 results in a crystallographic change of the material that leads to the formation of biphasic mixtures having various HA/TCP ratios depending on the starting Ca/P molar ratios. With a starting Ca/P ratio higher than that of s-HA, free CaO will separate out as the secondary phase at >900°C. The thermal stability of

HA prepared at 25°C as a function of the mixing Ca/P molar ratio is explained by Osaka *et al.* [48] in the form of a pseudo phase diagram shown in Fig. **2**.

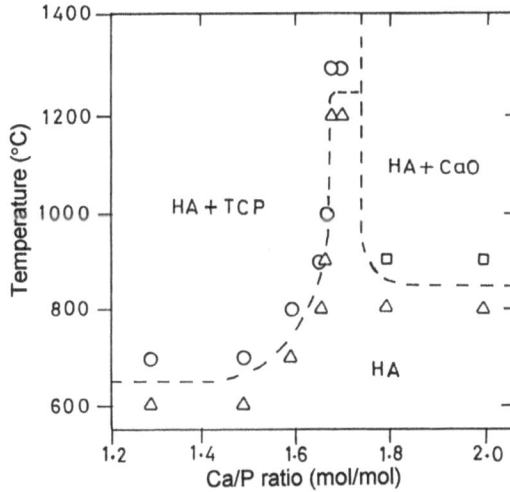

Figure 2: A pseudo phase diagram showing thermal stability of HA prepared at 25°C as a function of the mixing molar ratio Ca/P [48].

According to this, the temperature for appearance of TCP phase increases from 650 to 800°C as the Ca/P ratio is increased from 1.5 to 1.67 and CaO is appeared for Ca/P = >1.75 above 900°C. The most thermally stable HA, which is stable up to ~1250°C was formed from Ca:P mixing ratios ranging from 1.67 to 1.75. These findings correlate well with the results of other studies [40, 49, 55, 57]. Many investigators showed that pure s-HA is stable without any decomposition up to about 1300°C. However, above 1000°C, HA loses its OH⁻ ions gradually by the phenomena called dehydroxylation. At ~1300°C, the loss is complete and each HA molecule loses one H_2O molecule (14, 15) [47, 49, 55, 85-88].

$$2OH \text{--------} \rightarrow O + H_2O \uparrow \tag{14}$$

$$Ca_{10}(PO_4)_6(OH)_2 \text{--------} \rightarrow Ca_{10}(PO_4)_6O \, \square + H_2O \tag{15}$$

This hydroxyl ion deficient product with a vacancy ($\square$) is known as oxyhydroxy apatite or oxyapatite (OHA). One of the positions which was occupied by OH groups in HA unit cell, is now occupied by an oxygen atom and the other is vacant. The loss of OH⁻ ions from the HA lattice is confirmed by cell contraction observed by XRD and continuous reduction in the intensity of OH absorption bands in the IR spectra above 1000°C with its final disappearance around 1300-1350°C [47, 85, 89].

On further heating to temperatures in the range of 1300 - 1500°C, the HA or OHA decomposes in to TCP, TTCP and some times calcium pyrophosphate (16, 17) [85, 88, 89]. The TCP formed by decomposition of HA at high temperatures is in the form of α'-phase which later transforms into α-phase when the temperature is reduced.

$$Ca_{10}(PO_4)_6\,O\ \text{-------}\rightarrow 2Ca_3(PO_4)_2 + Ca_4P_2O_9 \qquad\qquad (16)$$

$$2\,Ca_{10}(PO_4)_6\,O\ \text{-------}\rightarrow 2Ca_3(PO_4)_2 + Ca_2P_2O_7 + 3\,Ca_4P_2O_9 \qquad\qquad (17)$$

The decomposition of HA is greatly influenced by the sintering atmosphere. It has been found that in air and vacuum the dehydroxylation of HA occurs in the temperature range 850-900°C, while in moisture there was no sign of dehydroxylation or decomposition at the highest sintering temperature of 1350°C [8, 30]. The decomposition of HA was observed at 1300°C in air while at as low as 1000°C in vacuum. Studies on thermal stability of HA at high temperatures reveals that pure HA can be conventionally sintered over a wide temperature range from 1000 to 1400°C depending upon the powder preparation method, its particle size, surface area, morphology *etc.* However, the optimum sintering temperature reported lies in the range of 1200-1300°C in ambient atmosphere while densification is better >1300°C under a moist environment. Under vacuum, densification will be largely retarded due to decomposition at as low as 1000°C. Further, poor densification of HA under moisture below 1200°C as compared to those of HA sintered in air show the limited sinterability of HA as compared to OHA. Another important factor in sintering of HA is the exaggerated grain growth at high temperatures >1300°C which leads to poor mechanical properties [4, 88, 90].

The thermal stability and densification of HA will also be greatly influenced by the additions of oxides or non-oxides either in the particulate [91, 92] or fiber [93] forms. They are added with the aim to improve the mechanical properties of the HA matrix. In order to achieve high densification of HA without thermal decomposition (with or without the presence of additives) with controlled grain size and optimum mechanical properties, hot pressing or HIPing under sufficient moisture content has been followed [81-83, 94]. There were attempts on the densification of HA at lower temperatures through liquid phase sintering using lithium, magnesium and sodium compounds as well as by phosphate and silicate glass additions [4]. The densification in the initial stages of sintering was found to be enhanced by liquid phase but resulted in larger grain growth leading to a lowering of mechanical properties. Fang *et al.* [95] fabricated HA ceramics having 97% theoretical density, improved microstructure and enhanced mechanical strength by a microwave sintering, which is also aided in saving the time and energy required for sintering.

3.4. Characteristic Properties of HA

The interesting characteristic properties of HA are: (a) physical properties like density, crystal structure, colour *etc.* (b) mechanical properties like compressive and bending strength, fracture toughness, hardness *etc.* (c) thermal properties like thermal expansion coefficient, melting point, thermal conductivity *etc.* and (d) chemical and biological properties like corrosion resistance, solubility, bioactivity *etc.* Some of the important properties of medical grade HA is listed in Table **1**, while typical characteristic properties of HA are compared with those of the human bone and other bioceramic materials as is shown in Table **2**. Various properties of HA are discussed in detail elsewhere [4, 19-30, 37, 96], and hence the same are not mentioned here. However, a brief discussion on the solubility, biodegradation, precipitation and bone bonding phenomena of HA and other calcium phosphate ceramics are presented considering its relevance for biomedical applications.

The solubility of HA is significant from the point of view of its stability under physiological conditions, its resorption and calcification. A thorough literature is available on the study of dissolution of HA, and is summarised by Narasaraju *et al.* [37]. HA being a salt of weak acid, may undergo hydrolysis in aqueous solutions (18, 19).

$$Ca_{10}(PO_4)_6\ (OH)_2 + 6H_2O \text{ ------} \rightarrow 4Ca_2(HPO_4)\ (OH)_2 + 2Ca^{2+} + 2HPO_4^{2-} \qquad \textbf{(18)}$$

$$4Ca_2(HPO_4)\ (OH)_2 \text{ -------} \rightarrow 8Ca^{2+} + 4HPO_4^{2-} + 8OH^- \qquad \textbf{(19)}$$

Table 1: Properties of medical grade HA [19-21, 96]

Property	Typical value	Remarks
Density	3.16 g cm^{-3}	Theoretical density, HA is always porous
Compressive strength	$100 - 200 \text{ MN m}^{-2}$	Dependent on porosity
Bend strength	$< 100 \text{ MN m}^{-2}$	Dependent on porosity
Fracture toughness	$< 1 \text{ MPa. m}^{1/2}$	Similar to window glass
Young's modulus	100 GN m^{-2} max.	Dependent on porosity
Hardness	500 HV	Similar to window glass
Thermal expansion coefficient	$11 \times 10^{-6} \text{ K}^{-1}$	-
Melting point	1650°C	Decomposition: HA should not be sintered at >1350°C
Corrosion resistance	Bioactive, Osseoconductive	Interaction in body environment
Colour	Various (white, bluish)	Dependent on raw material and process parameters

Table 2: Characteristic properties of some bioceramic materials as compared to skeletal tissues [1-12, 19-22, 97]

Properties	HA	Al_2O_3	ZrO_2 (PSZ)	Cortical Bone	Cancellous Bone	Enamel
Density (g cm^{-3})	3.16	3.98	6.0	1.5-2.2		2.9-3.0
Compressive strength (MPa)	270-900	4500	1800	100-300	2-23	250-400
Flexural strength (MPa)	80-250	380-600	800-1200	50-200	10-20	-
Tensile strength (MPa)	90-120	250	400	20-150	1	70
Hardness (Vickers)(GPa)	3-7	20	150	0.4-0.7	-	3.4-3.7
Young's modulus (GPa)	35-120	380-420	200	7-30	0.5	40-84
Fracture toughness; K_{IC} $(\text{MPa.m}^{1/2})$	0.6-1.2	3-6	7-15	2-12	-	-
Thermal conductivity $(\text{W m}^{-1} \text{°C}^{-1})$	1.3	35	1.7-3.5	0.6	-	0.9
Thermal Expansion Coefficient $(\times 10^{-6} \text{°C}^{-1})$	11-14	8-11	9-11	-	-	11

In vitro dissolution of HA depends on the type and concentration of the buffered or unbuffered solutions, pH of the solution, degree of saturation of the solution, solid-solution ratio, the duration of suspension in the solutions, the composition (secondary phases) and crystallinity of HA as well as, the defect structure and the micro and macroporosities of the product [22, 98, 99]. The s-HA is found more stable and the order of solubility for various calcium phosphates decreases in the order: TTCP > α-TCP > OHA > β-TCP > CDHA > S-HA. *In vitro* stability of pure HA, β-TCP and HA - β-TCP biphasic calcium phosphate (BCP) ceramics in deionised water and in physiological solutions investigated by Kohri *et al.* [100] has shown that pure β-TCP and β-TCP in BCP ceramics gradually converted to needle like crystals of HA, whereas pure HA showed no change. Thus a BCP ceramic with predetermined HA to β-TCP ratio will have the advantage of controlled biodegradation rates in clinical applications. Parallel to the above study, the *in vivo* bio-degradation of calcium phosphates investigated extensively [101,102] has shown that the phenomena of biodegradation is complex and contradicting. Generally s-HA is nearly non-biodegradable and the biodegradability of various calcium phosphates follows the order: β-TCP > rhenanite $(CaNaPO_4)$ > Mg-whitlockite $\{Ca_{2.5} Mg_{0.5} (PO_4)_2\}$ > HA.

The phenomena of calcification involves the orderly precipitation of HA within the organic matrix of bone. The formation of biological carbonate apatite (HCA) on the surface of implanted synthetic calcium phosphate ceramics leading to bonding with bone follows a sequence of chemical reactions at the implant tissue-interface, depending upon the surface chemistry of HA [22]. Radin *et al.* [103] analysed reactions of several single phase calcium phosphate ceramics upon immersion into a simulated physiological solution (SPS) and minimum time for measurable precipitation of HCA was found to increase in the order: not-well-crystallised HA < well-crystallised HA < α-TCP, TTCP < β-TCP. The rate of apatite mineral formation on the surface of the implant corresponds to the dissolution rate as well as to the precipitation rate of calcium phosphates *in vitro* [98-103] and thus the bonding mechanisms of synthetic HA through the formation of HCA crystals on the implant surface can be explained through dissolution-precipitation phenomena [5, 8, 22].

4. SYNTHESIS, PROCESSING, CONSOLIDATION AND CHARACTERI-SATION OF BIOCERAMICS

Bioceramics are generally produced by following the same synthesis, processing, consolidation and characterization methods that are used for high quality engineering ceramics. The properties of interest for a bioceramic are achieved through a systematic process control and characterization procedures. As in the production of any engineering material, the processing of a bioceramic involves different stages, that includes raw material selection or synthesis, processing, shaping, sintering, post treatment (finishing *etc.*) and a final inspection. The materials at raw material stage, at intermediate stage of processing as well as at the final finished stage undergoes various physical, chemical and mechanical characterizations similar to a engineering material. In addition, the material has to be evaluated for various biocompatibility tests through *in vitro* and *in vivo* studies.

In this section, case studies on synthesis and characterization of hydroxyapatite (HA) based ceramic powders, colloidal processing and shaping of hydroxyapatite and alumina are discussed.

4.1. Synthesis and Characterization of Hydroxyapatite (HA) Ceramics

Studies have been carried out on solid state synthesis and thermal stability of HA and HA - β-TCP biphasic powders followed by sintering and characterization of HA + β-TCP composites [4, 55, 89]. Powders of pure β-TCP, HA and a biphasic mixture of HA+β-TCP are prepared by solid state reaction between two commercially available calcium-based precursors namely, tricalcium phosphate (TCP) and calcium hydroxide [$Ca(OH)_2$]. These reactants mixed in the molar ratios ranging from 3:0 to 3:4 (designated T0 to T4) in deionized water, milled and slip-cast into discs are heat treated in the temperature range of 600°C to 1250°C. The products formed are characterized by X-ray diffraction (XRD) and infra red spectroscopic techniques for identification of phases formed and functional groups present in them. While tricalcium phosphate and calcium hydroxide taken in the molar ratio of 3:2 and 3:3 resulted in pure HA on heat treating at 1000°C for 8 h, the 3:1 and 3:1.5 molar ratio compositions resulted in a biphasic mixture of HA+β-TCP for similar heat treatments. Heat treatment of 3:4 molar ratio composition of tricalcium phosphate and calcium hydroxide at 1000°C yielded HA with free CaO as the additional secondary phase. On heat treatment at higher temperatures (1150 and 1250°C) for shorter duration (2 h), the products obtained from T0 and T2 are same as that obtained at 1000°C (pure β-TCP and pure HA). On the other hand, in the case of T1, T1.5, T3 and T4, the products have less percentages of HA with the formation of β-TCP (in the case of T1 and T1.5) or CaO (in the case of T3 and T4) as secondary phases.

From these results, it is concluded that heat treatment at the temperature of 1000°C/8 h for the as received, wet milled and slip cast powders as well as its mixtures with $Ca(OH)_2$ in various proportions transforms them into stable products consisting of either pure β-TCP, pure HA or mixtures of HA+β-TCP or HA+CaO depending upon the extent (*x*) of $Ca(OH)_2$ (0<x<4) used for the reaction. High percentage of $Ca(OH)_2$ addition (*e.g.* in T4) leads to the presence of excess $CaO/Ca(OH)_2$ as the secondary phase which is harmful for sintering and achieving good mechanical properties for the HA formed. While pure HA formed from T2 is stable even up to temperature of 1250°C, the HA formed from other compositions are unstable and decomposed resulting in increased amount of β-TCP (T1 and T1.5 cases) or CaO (T3 and T4 cases). The method is therefore suitable for the preparation of pure stoichiometric β-TCP, HA as well as their biphasic (HA+β-TCP) composite powder mixtures [55].

Since HA has poor mechanical properties, it has limitation in its usage for low load bearing applications. Increasing the toughness of HA by reinforcing with a second phase material is a subject of interest for many researchers. However, the phase stability and sintering of HA phase is found poor in presence of additives, since it decomposes to β-TCP and other phases. Hence, synthesis of HA in the presence of oxide additives like ZrO_2 and Al_2O_3 and their thermal stability as well as sinterability is studied [91, 92]. Hydroxyapatite (HA) is synthesised in the presence of 10–30 wt.% of m-ZrO_2 by solid state reaction between TCP and $Ca(OH)_2$ at 1000°C for 8h. The m-ZrO_2 is partly converted into t-ZrO_2 by partial consumption of CaO which in turn resulted in a mixture of β-TCP and HA. On sintering, these HA–β-TCP–ZrO_2 composite powders at 1100–1400°C for 2h, HA is further decomposed into β-TCP and CaO. The CaO so produced reacts further with m-ZrO_2/t-ZrO_2 generating a mixture of t-ZrO_2 and $CaZrO_3$ in different proportions. These various phases formed interfere with the sinterability of the composites due to their differential shrinkages leading to a overall reduced density as compared to that of pure HA. The composites show a T-onset of decomposition at around 1150°C and a 40% HA yield is obtained at the highest sintering temperature of 1400°C. These results show that higher values for percentage HA yield, the T-onset of decomposition and the densification at various temperatures are achieved for ZrO_2-doped HA composites in the present study as compared to the values reported in the literature. This leads to a very significant observation that, high-temperature solid state reaction of the precursors processed through wet milling route results in more dense and more thermally stable HA–ZrO$_2$ composite products than those obtained by other (*e.g.* coprecipitation) routes [92]. Similar studies carried out with Al$_2$O$_3$ additions has resulted in composite products with considerable amount of HA phase retained along with β-TCP and calcium aluminate as the secondary phases [91].

Recently, synthesis of nano sized HA and HA-βTCP biphasic ceramic powders is carried out through solution combustion route [104]. The effect of various fuels and their combinations on the reaction condition and characteristics of the powder produced are investigated. The powders are characterized for phase and functional group, crystallite and particle size, agglomerate size, morphology, surface area and surface charge. From the systematic study on the solution combustion synthesis it could be concluded that the type of fuel, their combinations, oxidizer to fuel ratio and the furnace temperature have a determining effect on the type and characteristics of combustion reaction and physical properties of the powder formed like phase, crystallinity, particle size, surface area, dispersability *etc.* While the reaction is incomplete with urea alone as fuel at the experimental conditions considered, glycine give a product with some carbonaceous matter left as residue which required further calcination for its removal. On the contrary, by using urea and glycine fuel mixtures well crystallised HA and BCP powders having crystallite size in the range of 10-20 nm, particle size of 100-200 nm, surface area 10-20 m^2/g and elongated morphology could be prepared up to 50 g batches using simple laboratory mantle heater at considerably lower temperature of about 300°C. The nano HA powder thus produced by this method is also considered as a promising candidate in targeted drug delivery formulation for osteoporosis treatment for rats. This proves that a proper combination of fuel may reduce the preheating temperature and hence lower is the energy consumption in producing technologically important materials like HA and BCP through a single step solution combustion method.

4.2. Colloidal Processing of Al$_2$O$_3$ and HA

The processing and shaping of bioceramics, specially involving sub-micrometer sized powders requires a high degree of optimisation to achieve the desired microstructural characteristics in the green body. The fine ceramic particles are generally agglomerated in the as prepared condition. The hard agglomerates are broken down into smaller fragments or individual primary particles by mechanical milling. The soft agglomerates formed between the fine particles due to physical force of attraction (van der Waals force) become the crucial defect center leading to inhomogeneous green microstructure. This affects the sintered density, microstructure and final properties. Thus, the main goal in the processing of fine ceramic powders is to avoid or minimize these agglomerates.

One novel approach of avoiding or minimizing these agglomeration, is the colloidal dispersion of fine ceramic powders in an aqueous or non-aqueous liquid medium and their stabilization before consolidation. The principle of colloidal processing lies in the fact that, generally the metallic oxide surfaces like Al_2O_3, SiO_2 *etc.* when in contact with water get, hydrolysed by chemisorption leading to the free OH^- groups getting appended to the particle surfaces. These hydrolysed surfaces undergo acid/base reactions depending on the pH of the medium and develop positive or negative surface charges.

$$M\text{-}OH\ (s) + H^+\ (aq) \rightarrow M\text{-}OH_2^+\ (s) \hspace{3cm} \text{(acidic medium)}$$

$$M\text{-}OH\ (s) + OH^-\ (aq) \rightarrow M\text{-}O^-\ (s) + H_2O\ (l) \hspace{2cm} \text{(basic medium)}$$

Where M = Metal Atom Like Al, Si *etc.*

Other types of surface reactions also occur depending upon the surface groups, like amine (-NH) in Si_3N_4, phosphates in hydroxyapatite *etc.* The surface of the powder particles can also be modified by the adsorption of specific ionic groups (inorganic or organic) in the form of electrolytes or polyelectrolytes which are known as dispersing agents [4, 72, 105-109].

Such electrostatic surface charges developed due to acid-base reactions or because of adsorbed ionic groups promote repulsion between the particles by overcoming the van der Waals attractive forces. This result in breaking of spontaneously formed soft agglomerates into fine individual particles leading to improved dispersion. Thus, the dispersion and stabilisation of a ceramic powder suspension is controlled by the surface characteristics of the particles and through the balance between van der Waals attractive forces and the electrostatic repulsive forces developed on the surface of the particles either by adjustment of pH of the medium or by the addition of electrolytes. The charged particles adsorb oppositely charged ions called counter ions forming a double layer of charges called as electrical double layer. A potential difference is developed between the charged particle and the adsorbed layer (stern layer) which is called as zeta potential in practical situations. The zeta potential will be indicated by negative value when particles are negatively charged and the positively charged particles give positive value for the zeta potential. Depending upon the polarity of the surface charge of the particles at different pH of the medium (acidic to basic range) the zeta potential may change from negative to positive or *vice versa*. The point at which the zeta potential is zero is referred to as the iso-electric point (IEP) wherein the number of positive charges equal to number of negative charges.

The zeta potential, which expresses the potential due to the net effective charge on the surface of particles is an effective diagnostic tool in optimising the dispersion behaviour of a ceramic slurry. With other factors being constant, a higher zeta potential indicates an increased electrostatic repulsion between two neighbouring particles leading to their separation and thereby to improved dispersion [4, 106-110]. At lower ZP and near IEP the lower surface charges lead to decreased repulsion and hence agglomeration of particles due to van der Waals force of attraction. The Zeta potential (ZP) values measured as a function of pH for Al_2O_3 and hydroxyapatite powders dispersed in deionised water with and without dispersant are depicted in Figs. **3** and **4** respectively. The alumina powder has high ZP in the pH region of 2 to 4 (> 40mv) and at pH 11 (> -35mv) showing these as the conditions of optimum dispersion, while the iso-electric point (IEP) is found at ~7.9 pH, which represents the region of high agglomeration. On the contrary by adding 0.2% of dispersing agent, the Zeta potential increased drastically towards negative value in the wide pH range with a shift in the IEP to a pH value of about 3. This shows adsorption of negatively charged polyelectrolyte on to the slightly negatively charged alumina powder resulting in higher surface charge and hence increased dispersability.

The zeta potential curves for nano hydroxyapatite particles in Fig. **4** show that in the absence of dispersant, the zeta potential values fluctuate in the range of -10 to +10 in the entire pH range. These values indicate that nano HA particles would be highly agglomerated due to the lower surface repulsion forces which are unable to overcome van der Waals attractive forces. On addition of a dispersant, the surface of the particles become more negatively charged and zeta potential values are drastically increased to high values of -30 to -50 in the pH range 9 – 11. This favors de-agglomeration of the nano particles by breaking down the soft agglomerates leading to good dispersion.

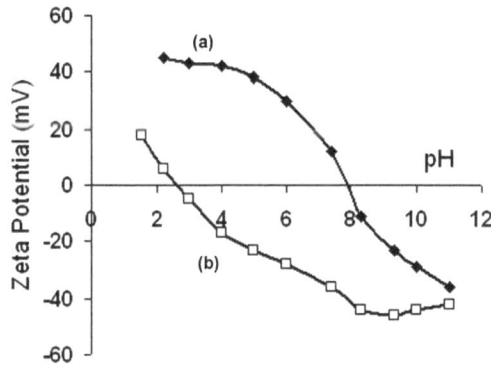

Figure 3: Zeta potential for Al_2O_3 in deionised water (a) without and (b) with dispersant.

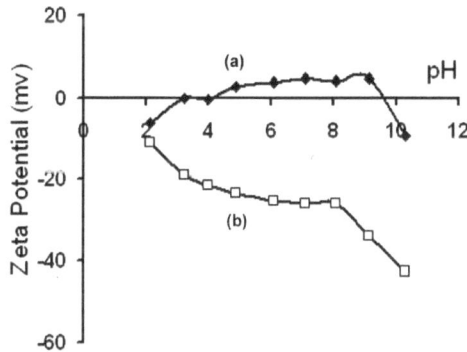

Figure 4: Zeta potential for nano HA in deionised water (a) without and (b) with dispersant.

Particle size distribution, sedimentation behaviour of the particles under suspension, viscosity and rheological behaviour of the ceramic slurry are other important characteristics that are influenced by the dispersion quality of ceramic powders in a liquid medium. When the particles are agglomerated, they sediment at faster rate due to high mass and they occupy large volume resulting in higher sediment height. On the other hand, when the particles are well dispersed the individual particles sediment at relatively slower rate and occupy lower volume leading to lower sediment height due to better packing of the particles. The detailed discussions on Al_2O_3 and HA powders are discussed elsewhere [72, 108, 109]. The results very much correlate with the zeta potential measurement. The powders show lower sedimentation heights at pH of higher zeta potential and higher sedimentation heights at the pH of lower zeta potential values and near the IEP. The addition of dispersant, which result in enhanced repulsion between the particles as indicated by higher zeta potential lead to minimum sedimentation height. The pH and/or dispersants resulting in lower sedimentation height are the conditions of optimum dispersion of fine particles in the liquid medium.

In continuation with sedimentation studies, the dispersion behaviour of the above powders is well explained by the viscosity and rheological measurements. The viscosity is high at the condition of agglomeration as part of the liquid is entrapped between the particles and not available for wetting the particle surfaces. However, in the case of well dispersed condition, the viscosity is low as maximum liquid is available for wetting the particles. The viscosity *versus* pH curve for alumina slurry presented in Fig. **5** shows that the dispersion is better in

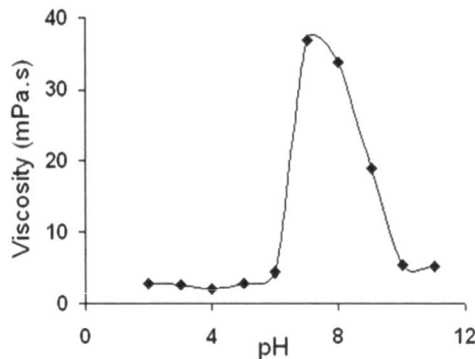

Figure 5: Viscosity of Al_2O_3 slurry as a function of pH.

two pH ranges of 2 to 5 and 10 to 11, attributable to high zeta potential values. On the other hand, very high viscosity values in the pH region of 6 to 10 could be related to very high agglomeration resulting from lower zeta potential values around its IEP. In a similar manner, the viscosity of Al_2O_3 and HA slip measured at different shear rates are affected by the pH of the medium as well as the dispersant and the results are in correlation with those of zeta potential and sedimentation studies. The addition of an organic dispersant, ammonium polymethacrylate resulted in considerable decrease in the viscosity of HA slurry as shown in Fig. **6** [72]. From this study, it is concluded that 0.2-0.3% of dispersant is the optimum amount required for achieving the optimum dispersion of alumina, whereas 2-4% of dispersant gave near Newtonian flow for 50 wt.% HA slurry [72, 108, 111].

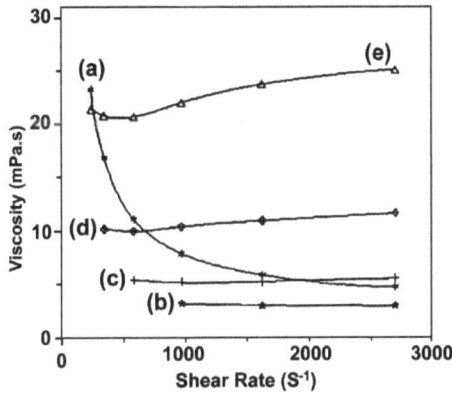

Figure 6: Viscosity of HA slurry as a function of shear rate (a) without dispersant at 35 wt% solid loading (b-e) with 3 wt% dispersant at 35, 45, 50 and 55 wt% solid loading.

4.3. Slip Casting of Bioceramics

Slip casting is one of the traditional shaping or forming method for clay based ceramic materials originated between 1700 and 1740 AD. Basically, slip casting involves four stages. *viz*: (1) mould preparation (2) slip preparation (3) slip casting and (4) green body removal from the mould.

Mould is a negative replica of the shape to be prepared by slip casting. Generally, a mould is prepared by pouring plaster of paris slurry around the master die. After hardening (setting), the mould is separated and dried. In the second step, highly solid loaded slip is prepared by dispersing the fine ceramic powder particles in a liquid medium. This is followed by pouring the slip into the mould. During this casting step liquid is absorbed into the mould through fine pores present in them by capillary action and solid body is developed which has the shape corresponding to the inner contour of the mould. This is called as the green body and is removed from the mould, dried and then sintered at high temperatures to get the final product.

After 1910, when alumina, the first non-clay material is slip cast, the technique has been employed successfully for casting of various oxide ceramics like silica, magnesia, zirconia, calcia etc as well as for non-oxide materials. Slip casting has emerged as one of the forming technique for large scale fabrication of both monolithic as well as composite advanced ceramic components of either very simple or complicated shapes. During the last two decades, colloidal dispersion of advanced ceramic powder particles in a liquid medium has been practiced before their consolidation by slip casting. The colloidal dispersion of fine powders is very much determining in case of non-clay (oxide and non-oxide) ceramic materials in order to achieve highly solid loaded homogeneous slip with minimum viscosity and near newtonian flow behaviour which in turn leads to dense green compacts with minimum (in number/size) defects on slip casting [109-112]. Recently, the colloidal processing followed by slip casting has been employed for variety of dense and porous shapes from HA and other calcium phosphate powders for biomedical applications [70-73].

By employing the optimum slip conditions that results uniform dispersion, a high solid loaded (52-55 Vol.%) Al_2O_3 slip is prepared in deionised water, milled in polythene bottles using alumina balls as milling media for about 12-16 h and cast in plaster moulds of various shapes. The green densities achieved in the range 53 to 58% of theoretical value correlate very well with the results available in the literature for slip cast alumina products. The slip cast Al_2O_3 samples sintered to 98±1 % density in the temperature range of 1500 to 1600°C for 2 h show flexural strength values of 320 ± 20 MPa. The detailed discussion of these studies is presented elsewhere [108, 109, 111].

Similarly slip casting of optimised HA slip is carried out to result in green bodies having a density of ~ 50 % of theoretical value. On sintering these slip cast HA samples posses a density of ~ 93 %, porosity of 5±0.5% and flexural strength value of 55±5MPa. The microporosity in the range of 1-2 μm is expected to be beneficial in biomedical applications. The details of the studies carried out on slurry optimisation, slip casting, sintering and characterisation of HA are presented elsewhere [72].

4.4. Gel Casting of Bioceramics

In contrast to slip casting, gel casting is a direct consolidation technique in which fluid slurry is transformed into homogeneous rigid body without liquid removal, in a nonporous mold. The slurry is prepared by mixing a fine ceramic powder with deionised water containing monomer (*viz.* Acrylamide), cross linking agent, initiator and catalyst. After preparation of homogenous slurry, it is poured into a nonporous mold made of aluminium, wax or rubber. Inside the mould the fluid slurry transforms into a rigid body by *in situ* polymerization of the monomer. After completion of polymerization, the samples are removed from the mould and dried. Because of the organic polymer acting as binder, the green body exhibit sufficient strength so that they can be easily handled as well as machining can be employed to achieve required complicated shapes or surface features. The organic binder will be removed by slow heating in the temperature range of 300-600°C (binder burnout stage) and then sintered at high temperatures to obtain dense body [113-116].

Gel casting has been developed as a near net shaping technique to produce complex shapes, thin sections as well as micro featured components. The suspension needs to be of high concentration (> 50% in volume) so that a sufficiently high green density is achieved and a closer dimensional control is possible. In a study gel casting is employed successfully for Al_2O_3 and calcium phosphate ceramics to produce variously shaped articles. The physical, micro-structural and mechanical characterization of the gel cast samples are carried out. The properties are comparable to those obtained for slip cast samples [114]. Fig. **7** shows the photographs of gel cast calcium phosphate samples having complex surface features and macrostructures.

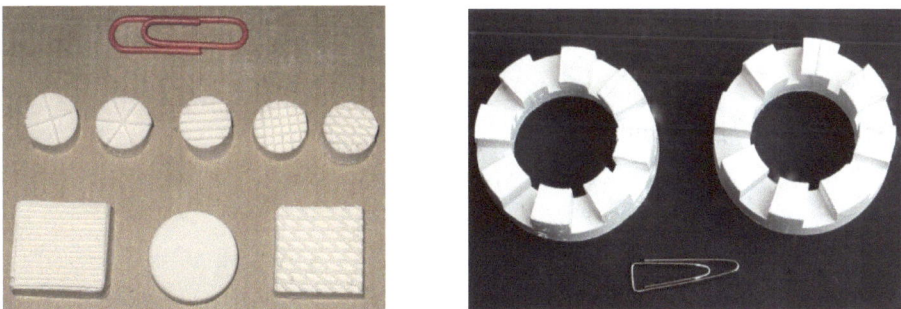

Figure 7: Gel cast calcium phosphate samples.

4.5. Solid Freeform Fabrication – Mouldless casting

Solid Freeform Fabrication (SFF) is an emerging, novel processing method being practiced for the past one decade to yield complex shaped 3-dimensional objects directly from a computer-created representation. SFF involves (a) computer aided designing (CAD) of the component image (b) layer-by-layer slicing of the 3-dimensional image into 2-dimensional representation and transfer these data into 3-dimensional movement manipulative device, and (c) layer-by-layer building of the object through 3-D movement manipulative device by using an, appropriate CAM software.

The layer-by-layer building of the ceramic components can be done through (a) stacking of green sheets made by tape casting (laminated object manufacturing), (b) solidification of free flowing powder or suspension spread uniformly in thin layers [Selective Laser Sintering (SLS), Stereolithography (SL), Drop-on-Demand printing] and (c) direct deposition of highly solid loaded suspension or semi solid mass in X, Y direction and building in Z direction [three dimensional printing, direct ink jet printing, robocasting, fused deposition of ceramics (FDC)]. In short SFF is a moldless manufacturing technique, which brings the concept into reality. The review of various SFF techniques and the adoption of some of them for the calcium phosphate and other biomaterials can be obtained from the literature [117-121].

The moldless casting or robocasting method practiced at National Aerospace Laboratories employs a computer controlled, deposition of highly solid loaded (55–58 Vol. %) ceramic slurries in sequential layers by extruding through a fine nozzle of 0.5 to 1 mm diameter. The slurry conditions are optimized through a colloidal stabilization techniques and the structural integrity of the deposited layer relies on the pseudoplastic rheology of the slurry. The experiments are conducted on alumina, calcium phosphate, alumina + hydroxyapatite and zirconia slurries. The mouldless cast samples are sintered to 85-98% density depending upon the material composition [111, 113, 119]. Fig. **8** shows the photograph of sintered calcium phosphate shapes prepared by moldless casting method. The structures having 3-dimensional interconnected porosity are expected to be useful as scaffolds in tissue engineering applications. Further study is in progress to optimise the fabrication and utilization of such structures.

4.6. Porous Ceramics

In contrast to dense ceramic products, those having porosities at various levels and architecture find applications in diversified fields ranging from catalysts to tissue engineering. The porosities may be in the nano, micro, meso as well as macro size range. The pores may be closed or open and their concentration may vary depending upon the application.

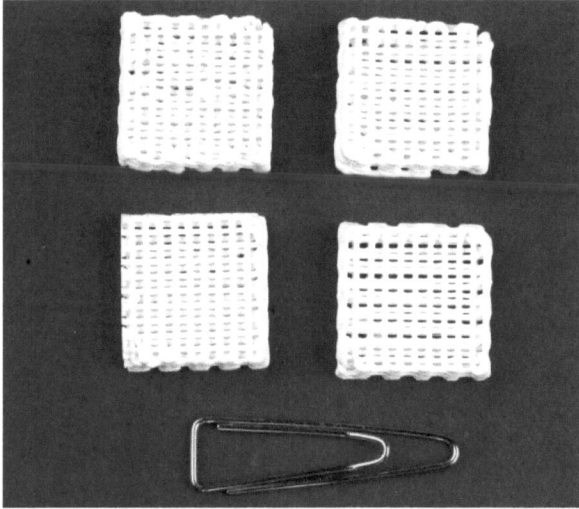

Figure 8: Sintered calcium phosphate shapes prepared by mouldless casting method.

For example, the scaffolds for tissue engineering requires multidimensional porosity in their structure *viz.*, micropores (diameter <10μm), mesopores (diameter 10μm – 100μm), macropores (diameter > 100μm) and straight channeled interconnected pore networks called as global pores having channel diameters in the range of 500 to 1000μm. Porous structure provides opportunity for in-growth of tissues while dense body gives a mechanical stability to the structure. The controlled pore architecture play a crucial role in giving biomaterial with improved adhesion, proliferation and differentiation of cells which leads to better osteoconductivity, bioactivity and mechanical properties. The high surface area obtained as the consequence of increased number of micropores is essential for oseoinduction. The concavities present on the walls of macropores resemble the geometric dependent of bone formation. Further, the dissolution of the surface causes supersaturation of calcium and phosphate ions which leads to their reprecipitation and formation of biological apatite. This property allows bone bonding with the bioceramics and influences its osteoinduction potential. The global pores facilitate the growth of tissues into the structure [120, 121]. The porous HA are also becoming attractive materials in controlled drug release applications [122].

A large number of methods are employed by researchers to fabricate the porous structure in advanced ceramic products which include use of porogenes or pore formers, foaming agents, replication methods as well as SFF techniques [111, 113, 123]. Fig. **9** shows the porous HA samples produced by foaming and slip casting method as well as the SEM micrograph showing details of the macro pore architecture of the sample. Fig. **10** shows the interconnected global pore networks produced by moldless casting method and the microporosity of the same sample. Thus, by using advanced shaping techniques, structures with multidimensional porosity can be prepared.

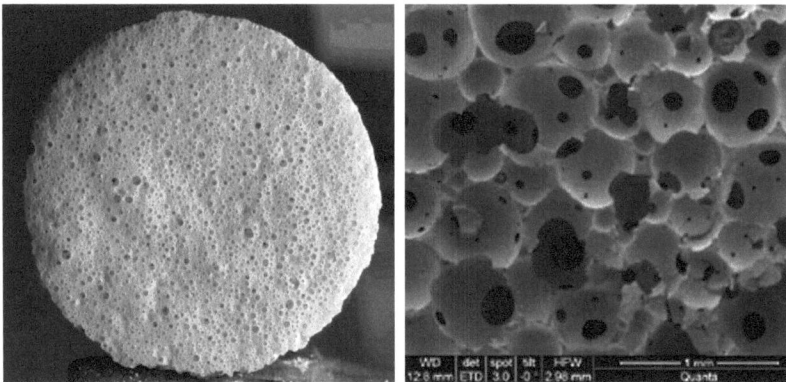

Figure 9: Porous HA disc (Left) and SEM of pore architecture (Right).

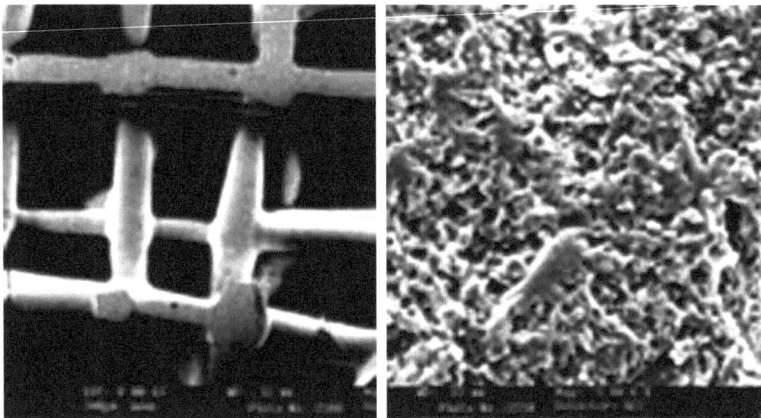

Figure 10: Global porosity (Left) and Microporosity (Right) in calcium phosphate.

CONCLUSIONS

Though the application area of biomaterials is highly specific and advanced, its production and utilization deals with almost all aspects of science and engineering including chemistry, physics, materials science, metallurgy, biomedical science, design engineering *etc.* As it is very hard to discuss about such an interdisciplinary subject, an attempt is made to explore some preliminary aspects of biomaterials specific to bioceramics. Though there is rapid increase in the list of materials to be used as a biomaterial in recent years, there still appears a vast opportunity for using a large number of conventional materials which can provide ample scope for a researcher to develop a biomaterial even to this date. In this chapter, a short review on bioceramics is presented followed by a discussion on the hydroxyapatite materials including its synthesis, processing and some physicochemical aspects. Brief discussion on the work carried out on synthesis, colloidal processing and shaping of HA, Al_2O_3 and HA-Al_2O_3 based ceramics is presented as case study.

A serious discussion on the science involved in processing and shaping of bioceramics in the fabrication of its components with high performance is required. The development of a technology for the fabrication of complex shaped components is possible only by a multidisciplinary approach. Tremendous scope exists for the research and development in the area of processing and shaping of bioceramic materials.

CONFLICT OF INTEREST

The contents presented in the chapter have been carefully written based on the review from the references cited and the results obtained from the investigations carried out by the author. Further, there is no conflict of interest with other people or organizations in respect of the present research work.

ACKNOWLEDGEMENTS

Author wishes to thank Director, CSIR-NAL and Head, Materials Science Division, CSIR-NAL for permitting to publish this chapter. Author is highly thankful to Dr. T. S. Kannan, for his sustained encouragement and the technical guidance given during some of the investigations carried out which has been quoted in this chapter. Thanks are also due to L. Mariappan and H.N. Roopa for their technical assistance and K. Venkateswaralu for his valuable contribution in preparation of this article.

REFERENCES

[1]   Williams DF. The Science and applications of Biomaterials. Advances in Materials Technology Monitor 1994; 1(2):1-10.
[2]   Dubok VA. Bioceramics – yesterday, today, tomorrow. Powder Metallurgy and Metal Ceramics 2000; 39(7-8):381-94.
[3]   Nomura N. Artificial organs: recent progress in metals and ceramics. J Artif Organs 2010; 13:10-2.
[4]   Ramachandra Rao R. Studies on Processing, Consolidation and Characterisation of Some Advanced Structural and Bio-Ceramic Materials. PhD dissertation. Mangalore University 2001.
[5]   Cao W, Hench LL. Bioactive Mateirals. Ceramics International 1996; 22(6):493-507.
[6]   Williams DF. Review, Tissue-biomaterial interactions. J Mat Sci 1997; 22(10):3421-45.
[7]   Hench LL, Wilson J. Introduction. In: McLaren M, Niesz DE, Eds. An introduction to bioceramics, Advanced series in Ceramics, Vol. 1. Singapore: World Scientific 1993; pp. 1-24.
[8]   Hench LL. Bioceramics: From concept to clinic. J Am Ceram Soc 1991; 74(7):1487-510.
[9]   Shirtliff VJ, Hench LL. Bioactive Materials for tissue engineering, regeneration and repair. J Mater Sci 2003; 38(23):4697-707.
[10]   Hulbert SF, Bokros JC, Hench LL, Wilson J, Heimke G. Ceramics in clinical applications: Past, present and future. In: Vincenzini P, Ed. High Tech. Ceramics. Amsterdam: Elsevier Science Pub 1987; pp. 189-213.
[11]   Hench LL, Wilson J, Eds. McLaren M, Niesz DE, Eds. in Chief. An introduction to Bioceramics, Advanced series in ceramics. Vol. 1. Singapore: World Scientific 1993; pp. 1-379.
[12]   Hench LL. Bioactive ceramics for skeletal repair. In: Birkby I, Ed. Ceramic technology international. London: Sterling publications Limited 1997; pp. 29-32.
[13]   Inamori K. Biological Applications. In: Somiya S, Ed. Advanced Technical Ceramics. Tokyo: Academic Press Inc 1984; pp. 209-22.
[14]   Phillips RW, Ed. Skinner's Science of Dental Materials, Ninth Edition. Bangalore, India: W. B. Saunders Company, Harcourt Brace Jovanovich, Inc., & Prism Books Pvt. Limited 1991.
[15]   Le Guehennec L, Layrolle P, Daculsi G. A Review of Bioceramics and Fibrin sealant. European Cells and Materials 2004; 8:1-11.
[16]   Best SM, Porter AE, Thian ES, Huang J Bioceramics: Past, Present and for the future. J European Ceramic Society 2008; 28(7):1319-27.
[17]   Stevens B, Yang Y, Mohandas A, Stucker B, Nguyen KT. A Review of materials, Fabrication Methods, and Strategies Used to Enhance Bone Regeneration in Engineered Bone Tissues. J Biomedical Mater Res Part B: Applied Biomaterials 2007; 573-82.

[18]   Oh S, Oh N, Appleford M, Ong JL. Bioceramics for Tissue Engineering Applications – A Review. Am J of Biochem Biotech 2006; 2(2):49-56.

[19]   Willmann G. Production of Bioceramics, Part I: Alumina ceramics, Part II: Hydroxyapatite ceramics. Interceram 1995; 44 (4/5) (Suppl 2.9.4):1-10.

[20]   Willmann G. Medical grade ceramics - what every engineer should know. Interceram 1998; **47** (1/2) (Supl 2.9.1.2.4):1-8.

[21]   Willmann G. Medical grade hydroxyapatite; state of the art. Brit Ceram trans 1996; 95(5):212-6.

[22]   LeGeros RZ, LeGeros JP. Dense hydroxyapatite. In: Hench LL, Wilson J, Eds. McLaren M, Niesz DE, Eds in Chief. An introduction to Bioceramics, Advanced Series in Ceramics Vol. 1. Singapore: World Scientific 1993; pp.139-80.

[23]   Uchida U, Nade SML, McCartney ER, Ching W. The use of ceramics for bone replacement, A comparative study of three different porous ceramics. The J Bone and Joint Surgery 1984; 66B(2):269-75.

[24]   Hench LL, Wilson J. Surface active Biomaterials. Science 1964; 226 (4675):630-6.

[25]   Lavernia C, Schoenung JM. Calcium Phosphate ceramics as Bone Substitutes. Am Ceram Soc Bull 1991; 70(1):95-100.

[26]   Jarcho M. Biomaterial aspects of calcium phosphates, properties and applications. Dent. Clin of North America 1986; 30(1):25-47.

[27]   Metsger DS, Driskell TD, Paulsrud JR. Tricalcium phosphate ceramic - a resorbable bone implant: review and current status. J Am Dental Association 1982; 105:1035-8.

[28]   Constantz BR, Ison IC, Fulmer MT, *et al.* Skeletal repair by *in situ* formation of the mineral phase of bone. Science 1995; 267(5205):1796-9.

[29]   Le Geros RZ, Le Geros JP. Calcium phosphate Bioceramics: Past, Present and Future. Key Eng Materials 2003; 240-242:3-10

[30]   de Groot K. Clinical Applications of Calcium Phosphate Biomaterials: A Review. Ceramics International 1993, 19(5):363-6.

[31]   Marquis P. Enhancing the strength of bioceramics. In: Birkby I, Ed. Ceramic technology International. Sterling publications Limited 1995; pp. 53-6.

[32]   Murugan R, Ramakrishna S. Effect of zirconia on the formation of calcium phosphate bioceramics under microwave irradiation. Mat Letters 2003; 58:230-4.

[33]   Soballe K, Hansen ES, Brockstedt-Rasmussen H, Bunger C. Hydroxyapatite coating converts fibrous tissue to bone around loaded implants. J Bone Joint Surg 1993; 75-B(2):270-8.

[34]   Jarcho M, Bolen CH, Thomas MB, Bobick J, Kay JF, Doremus RH. Hydroxyapatite synthesis and characterisation in dense polycrystalline form. J Mat Sci 1976; 11 (11):2027-35.

[35]   Peelen JGC, Rejda BV, de Groot K. Preparation and properties of sintered hydroxyapatite. Ceramergia Int 1980; 4:71-3.

[36]   Akao M, Aoki H, Kato K. Mechanical properties of sintered hydroxyapatite for prosthetic applications. J Mat Sci 1981; 16(3):71-3.

[37]   Narasaraju TSB, Phebe DE. Review, Some physico-chemical aspects of hydroxyapatite. J Mat Sci 1996; 31(1):1-21.

[38]   Brown PW. Phase relationships in the ternary system $CaO-P_2O_5-H_2O$ at 25°C. J Am Ceram Soc. 1992; 75(1):17-22.

[39]   Hayek E, Newsely H. Pentacalcium monohydroxy orthophosphate. Inorg Syn 1963; 7:63-5.

[40] Bako Z, Kotsis I. Composition of precipitated calcium phosphate ceramics. Ceramics International 1992; 18(6):373-8.

[41] Cuneyttas A, Korkusuz F, Timucin M, Akkas N. An investigation of the chemical synthesis and high temperature sintering behaviour of calcium hydroxyapatite (HA) and tricalcium phosphate (TCP) bioceramics. J Mat Sci Mat in medicine 1997; 8(2):91-6.

[42] Chaair H, Heughebaert J-C, Heughebaert M. Precipitation of stoichiometric apatitic tricalcium phosphate prepared by a continuous process. J Mat Chem 1995; 5(6):895-9.

[43] Correia RN, Magalhaes MCF, Marques PAAP, Senos AMR. Wet synthesis and characterisation of modified hydroxyapatite powders. J Mat Sci Mat in Medicine 1996; 7(8):501-5.

[44] Merry JC, Gibson IR, Best SM, Bonfield W. Synthesis and characterisation of carbonate hydroxyapatite. J Mat Sci Mat in medicine 1998; 9(12):779-83.

[45] Jha LJ, Best SM, Knowles JC, Rehman I, Santos JD, Bonfield W. Preparation and characterisation of fluoride-substituted apatites. J Mat Sci Mat in Medicine 1997; 8(4):185-91.

[46] Weng W, Baptista JL. A New synthesis of Hydroxyapatite. J Eur Ceram Soc 1997; 17(9):1151-6.

[47] Kijima T, Tsutsumi M. Preparation and thermal properties of dense polycrystalline oxyhydroxyapatite. J Am Ceram Soc 1979; 62(9-10):455-60.

[48] Osaka A, Miura Y, Takeuchi K, Asada M, Takahashi K. Calcium apatite prepared from calcium hydroxide and orthophosphoric acid. J Mat Sci Mat in Medicine 1991; 2(1):51-5.

[49] Slosarczyk A, Paluszkiewicz C, Gawlicki M, Paszkiewics Z. The FTIR spectroscopy and QXRD studies of calcium phosphate based materials produced from the powder precursors with different Ca/P ratios. Ceramics International 1997; 23(4):297-304.

[50] Dhondt CL, Verbeeck RMH, De Maeyer EAP. The growth of non-stoichiometric apatites using the constant composition method. J Mat Sci Mat in Medicine 1996; 7:201-5.

[51] Brown PW, Hocker N, Hoyle S. Variations in solution chemistry during the low temperature formation of hydroxyapatite. J Am Ceram Soc 1991; 74(8):1848-54.

[52] Tenhuisen KS, Brown PW. The kinetics of calcium deficient and stoichiometric hydroxyapatite formation from $CaHPO_4.2H_2O$ and $Ca_4(PO_4)_2O$. J Mat Sci Mat in Medicine 1996; 7(6):309-16.

[53] Fulmer MT, Brown PW. Hydrolysis of dicalcium phosphate dihydrate to hydroxyapatite. J Mat Sci Mat in Medicine 1998; 9(4):197-202.

[54] Nordstrom EG. Karlsson KH. Slip-Cast apatite Ceramics. Am Ceram Soc Bull 1990; 69(5):824-7.

[55] Ramachandra Rao R, Roopa HN, Kannan TS. Solid state synthesis and thermal stability of HAP and HAP-β-TCP composite ceramic powders. J Mat Sci Mat in Medicine 1997; 8(8):511-8.

[56] Yubao L, de Wijn J, Klein CPAT, de Meer SV, de Groot K. Preparation and characterisation of nanograde osteoapatite-like rod crystals. J Mat Sci Mat in Medicine 1994; 5(5):252-5.

[57] Yubao L, de Groot K, de Wijn J, Klein CPAT, de Meer SV. Morphology and composition of nanograde calcium phosphate needle like crystals formed by simple hydrothermal treatment. J Mat Sci Mat in Medicine 1994; 5(6/7):326-31.

[58] Suchanek W, Suda H, Yashima M, Kakihana M, Yoshimura M. Biocompatible whiskers with controlled morphology and stoichiometry. J Mat Res 1995; 10(3):521-9.

[59] Liu H. S, Chin TS, Lai LS, *et al.* Hydroxyapatite synthesised by a simplified hydrothermal method. Ceramics International 1997; 23(1):19-25.

[60] Hsu Y-S, Chang E, Liu H-S. Growth of phosphate coating on titanium substrate by hydrothermal process. Ceramics International 1998; 24(1):7-12.

[61] Gross KA, Chai CS, Kannangara GSK, Ben-nissan B, Hanley L. Thin hydroxyapatite coatings *via* sol-gel synthesis. J Mat Sci Mat in Medicine 1998; 9(12):839-43.

[62] Weng W, Baptista JL. Sol-gel derived porous hydroxyapatite coatings. J Mat Sci Mat in Medicine 1998; 9(3):159-63.

[63] Varma HK, Kalkura SN, Sivakumar R. Polymeric precursor route for the preparation of calcium phosphate compounds. Ceramics International 1998; 24(6):467-70.

[64] Toriyama M, Ravaglioli A, Krajewski A, Celotti G, Piancastelli A. Synthesis of hydroxyapatite-based powders by Mechano-Chemical method and their sintering. J Eur Ceram Soc 1996; 16(4):429-36.

[65] Shirkhanzadeh M. Direct formation of nanophase hydroxyapatite on cathodically polarised electrodes. J Mat Sci Mat in Medicine 1998; 9(2):67-72.

[66] Vijayaraghavan TV, Bensalem A. Electrodeposition of apatite coating on pure titanium and titanium alloys. J Mat Sci Letters 1994; 13(24):1782-5.

[67] Lerner E, Sarig S, Azoury R. Enhanced maturation of hydroxyapatite from aqueous solutions using microwave irradiation. J Mat Sci Mat in Medicine 1991; 2(3):138-41.

[68] Cui FZ, Luo ZS, Feng QL. Highly adhesive hydroxyapatite coatings on titanium alloy formed by ion beam assisted deposition. J Mat Sci Mat in Medicine 1997; 8(7):403-5.

[69] de With G, Vandijk HJA, Hattu N, Prijs K. Preparation, microstructure and mechanical properties of dense polcrystalline hydroxyapatite. J Mat Sci 1981; 16(6):1592-8.

[70] Toriyama M, Ravaglioli A, Krajewski A, Galassi C, Roncari E, Piancastelli A. Slip casting of mechanochemically synthesised hydroxyapatite. J Mat Sci 1995; 30(12):3216-21.

[71] Lelievre F, Bernache - Assollant D, Chartier T. Influence of powder characteristics on the rheological behaviour of hydroxyapatite slurrics. J Mat Sci Mat in Medicine 1996; 7(8):489-94.

[72] Ramachandra Rao R, Kannan TS. Dispersion and Slip Casting of Hydroxyapatite. J Am Ceram Soc 2001; 84(8):1710-6.

[73] Terpstra RA, Van der Heijde JCT, Swaanen P, Zhang X, Gubbels G. Slip Casting of Hydroxyapatite Ceramics. In: Duran P, Fernandez JF, Eds. Third Euro. Ceramics vol. 3. 1993; pp. 61-6.

[74] Li T, Lee J, Kobayashi T, Aoki H. Hydroxyapatite coating by dipping method and bone bonding strength. J Mat Sci Mat in Medicine 1996; 7(6):355-7.

[75] Liu D-M. Fabrication of hydroxyapatite ceramic with controlled porosity. J Mat Sci Mat in Medicine 1997; 8(4):227-32.

[76] Fabbri M, Celotti GC, Ravaglioli A. Hydroxyapatite - based porous aggregates: physico-chemical nature, structure, texture and architecture. Biomaterials 1995; 16(3):225-8.

[77] Roy DM, Linnehan SK. Hydroxyapatite formed from coral skeletal carbonate by hydrothermal exchange. Nature 1974; 247(5438):220-2.

[78] Arita IH, Castano VM, Wilkinson DS. Synthesis and processing of hydroxyapatite ceramic tapes with controlled porosity. J Mat Sci Mat in Medicine 1995; 6(1):19-23.

[79] Liu DM. Preparation and characterisation of porous hydroxyapatite bioceramic *via* a slip-casting route. Ceramics International 1998; 24(6):441-6.

[80]   Zuang H, Hon M. The effect of calcination temperature on the behaviour of HA powder for injection moulding. Ceramics International 1997; 23(5):383-7.

[81]   Takikawa K, Akao M. Fabrication of transparent hydroxyapatite and application to bone marrow derived cell hydroxyapatite interaction observation *in vivo*. J Mat Sci Mat in Medicine 1996; 7(7):439-45.

[82]   Ioku K, Somiya S, Yoshimura M. Dense/porous layered apatite ceramics prepared by HIP post-sintering. J Mat Sci Letters 1989; 8(10):1203-4.

[83]   Takagi M, Mochida M, Uchida N, Saito K, Uematsu K. Filter cake forming and hot isostatic pressing for TZP-dispersed hydroxyapatite composites. J Mat Sci Mat in Medicine 1992; 3(3):199-203.

[84]   Shaw JH, Best SM, Bonfield W, Marsh A, Cotton J. Study of the application of viscous plastic processing to hydroxyapatite. J Mat Sci Letters 1995; 14(15):1055-7.

[85]   Zhou J, Zhang X, Chen J, Zeng S, de Groot K. High temperature characteristics of synthetic hydroxyapatite. J Mat Sci Mat in Medicine 1993; 4(1):83-5.

[86]   Ishikawa K, Ducheyne P, Radin S. Determination of the Ca/P ratio in Calcium deficient hydroxyapatite using X-ray diffraction analysis. J Mat Sci Mat in Medicine 1993; 4(2):165-8.

[87]   Royer A, Viguie JC, Heughebaert M, Heughebaert JC. Stoichiometry of hydroxyapatite: influence on the flexural strength, J Mat Sci Mat in Medicine 1993; 4(1):76-82.

[88]   Van Landuyt P, Li F, Keustermans JP, Streydio JM, Delannay F, Munting E. The influence of high sintering temperatures on the mechanical properties of hydroxyapatite. J Mat Sci Mat in Medicine 1995; 6(1):8-13.

[89]   Ramachandra Rao R, Roopa HN, Kannan TS. Synthesis of pure HAP and HAP-β-TCP biphasic mixtures by solid state reaction. Trans. Ind Ceram Soc 1998; 57(3):73-6.

[90]   Tampieri A, Celotti G, Szontagh F, Landi E. Sintering and characterisation of HA and TCP bioceramics with control of their strength and phase purity. J Mat Sci Mat in Medicine 1997; 8(1):29-37.

[91]   Ramachandra Rao R, Kannan TS. Synthesis and sintering of hydroxyapatite in presence of oxide additives. Trans Ind Ceram Soc 1999; 58(3):64-8.

[92]   Ramachandra Rao R, Kannan TS. Synthesis and Sintering of Hydroxyapatite - Zirconia Composites. Materials Science & Engineering: C 2002; 20:187-93.

[93]   Ruys AJ, Milthorpe BK, Sorrell CC. Short-fibre-reinforced hydroxyapatite: Effects of processing on thermal stability. J Aust Ceram Soc 1993; 29(1/2):39-49.

[94]   Gautier S, Champion E, Bernache-Assollant D. Processing, Microstructure and toughness of Al_2O_3 platelet-reinforced hydroxyapatite. J Eur Ceram Soc 1997; 17(11):1361-9.

[95]   Fang Y, Agrawal DK, Roy DM, Roy R. Microwave sintering of hydroxyapatite ceramics. J Mat Res 1994; 9(1):180-7.

[96]   Willmann G. Material properties of hydroxyapatite ceramics. Interceram 1993; 42(4):206-8.

[97]   Zhang X-M. Processing and Mechanical properties of Nickel (Titanium)/Alumina and metal/hydroxyapatite composites. PhD dissertation. Eindhoven, Netherlands 1994.

[98]   Ducheyne P, Radin S, King L. The effect of calcium phosphate ceramic composition and structure on *in vitro* behaviour, I. Dissolution. J Biomed Mat Res 1993; 27:25-34.

[99]   Berger G, Gildenhaar R, Ploska U, Driessens FCM, Planell JA. Short-term dissolution behaviour of some calcium phosphate cements and ceramics. J Mat Sci Letters 1997; 16(15):1267-9.

[100] Kohri M, Miki K, Waite DE, Nakajima H, Okabe T. *In vitro* stability of biphasic calcium phosphate ceramics. Biomaterials 1993; 14(4):299-304.

[101] Klein CP, Driessen AA, de Groot K, van den Hooff A. Biodegradation behaviour of various calcium phosphate materials in bone tissue. J Biomed Mat Res 1983; 17:769-84.

[102] Ramselaar MMA, Driessens FCM, Kalk W, de Wijn JR, Van Mullem PJ. Biodegradation of four calcium phosphate ceramics, *in vivo* rates and tissue interactions. J Mat Sci Mat in Medicine 1991; 2(2):63-70.

[103] Radin SR, Ducheyne P. The effect of calcium phosphate ceramic composition and structure on *in vitro* behaviour, II. Precipitation. J Biomed Mater Res 1993; 27:35-45.

[104] Ramachandra Rao R, Mariappan L. Solution combustion synthesis and characterization of pure and biphasic nano hydroxyapatite, Unpublished work.

[105] Lange FF. Powder processing science and technology for increased reliability. J Am Ceram Soc 1989; 72(1):3-15.

[106] Sigmund WM, Bell NS, Bergstrom L. Novel Powder-processing methods for advanced ceramics. J Am Ceram Soc 2000; 83(7):1557-74.

[107] Lewis JA. Colloidal processing of ceramics. J Am Ceram Soc 2000; 83(10):2341-59.

[108] Ramachandra Rao R, Roopa HN. Colloidal processing and slip casting of Alumina-Zircon mixtures. In: Adhikari B, Banerjee H, Banthia AK, *et al.* Eds. ISAMAP2K4. Proceedings of the International symposium on advanced materials and processing; 2004; Materials Science Centre, IIT, Kharagpur, India 2004; pp. 1143-50.

[109] Ramachandra Rao R, Roopa HN, Mariappan L. Processing and slip casting of alumina-zircon mixtures to produce alumina-mullite-zirconia composites. In: INCCOM-7, Proceedings of the ISAMPE National conference on composites; 2008; National Aerospace Laboratories, Bangalore, India 2008; pp. 260-6.

[110] Ramachandra Rao R, Roopa HN, Kannan TS. Effect of pH on the dispersability of silicon carbide powders in aqueous media. Ceramics International 1999; 25(3):223-30.

[111] Ramachandra Rao R. Processing and shaping techniques for dense and porous ceramic products. In: Mondal B, Basu J, Sinha GP, Eds. Current research trends in investment casting. New Delhi: Allied publishers Pvt. Ltd. 2006; pp. 91-100.

[112] Ramachandra Rao R, Roopa HN, Shridhar Prasad K, Kannan TS. Forming of simple components of silicon carbide, silicon nitride and silicon by slip casting methods. Report No. NAL PDMT 9331, Materials Science Division, National Aerospace Laboratories, Bangalore 1993.

[113] Ramachandra Rao R, Mariappan L. A comparative study on conventional and mouldless casting techniques for alumina ceramics, Proc. International conference on trends in product life cycle, modeling, simulation and synthesis, PLMSS-2008, NAL, Bangalore, India 17-19th November 2008; pp. 247-51

[114] Ranjith Kumar JV, Roopa HN, Ramachandra Rao R. Colloidal processing and gel casting of alumina, Proc. International conference on advanced materials and composites, ICAMC 2007, NIIST, Trivandram, India Oct. 2007; pp. 24-2.

[115] Ometete OO, Janney MA, Nunn SD. Gel casting: From laboratory development toward industrial production. J Eur Ceram Soc 1997; 17(2-3):407-13.

[116] Young AC, Omatete OO, Janney MA, Menchhofer PA. Gel casting of alumina. J. Am. Ceram. Soc. 1991; 74(3):612-8.

[117] Halloran JW. Freeform fabrication of ceramics. Brit Ceram Trans 1999; 98(6):299-303.

[118] Cesarano III J, Segalman R, Albuquerqe NM, Calvert P. Robocasting – Provides Mouldless Fabrication from Slurry Deposition. Ceramic Industry 1998; 148(4):94-102.

[119] Ramachandra Rao R, Mariappan L. Robocasting (mouldless casting) – A novel shaping technique for ceramics. In: Advanced manufacturing technologies – An overview, A CSIR Network project, AMT COR 0021, CMERI, Durgapur, 17th Dec. 2005; pp.63-6.

[120] Li X, Li D, Lu B, Tang Y, Wang L, Wang Z. Design and fabrication of CAP scaffolds by indirect solid free form fabrication. Rapid Prototyping Journal 2005; 11(5):312-8.

[121] Cesarano III J, Dellinger JG, Saavedra MP, *et al.* Customization of Load-Bearing Hydroxyapatite Lattice Scaffolds. Int J Appl Ceram Technol 2005; 2(3):212-20.

[122] Palazzo B, Sidoti M.C, Roveri N, *et al.* Controlled drug delivery from porous hydroxyapatite grafts: An experimental and theoretical approach. Mat Sci and Engineering C 2005; 25:207-13.

[123] Ramachandra Rao R, Roopa HN, Mariappan L. Fabrication and characterization of multidimensional porous structures of Alumina-Hydroxyapatite. Extended abstract, 71st Annual session of the Indian Ceramic Society, Bangalore, India 9-11 January 2008; 44-5.

CHAPTER 2

Current Glass-Ceramic Systems Used in Dentistry

Anthony Johnson[1,*], Pannapa Sinthuprasirt[2], Hawa Fathi[1] and Sarah Pollington[1]

[1]Academic Unit of Restorative Dentistry, School of Clinical Dentistry, University of Sheffield, Claremont Crescent, Sheffield S10 2TA, UK and [2]Faculty of Dentistry, Thammasat University, Kong Luang, Pathumthani, Thailand 12121

Abstract: Dental ceramic restorations are essentially oxide based glass-ceramic systems. Glass-ceramics are employed in medicine and dentistry because they are relatively easy to process and have impressive mechanical properties. In addition, all-ceramic dental restorations are attractive for both dentists and patients because they have excellent aesthetics and their low thermal conductivity makes them comfortable in the mouth. In addition, the material is extremely durable and relatively easy to manufacture into customised units. The first ceramic to be used in dental restoration was dental porcelain. Introduced in the 1960s, this material has shown excellent aesthetics and biocompatibility, but its strength is only adequate for a limited range of applications. The development of advanced dental material technologies has recently led to the introduction of a range of all-ceramic restorations in dentistry. In this chapter, the authors have summarised the fundamental principles of glass-ceramic technology particularly it's use in dentistry and also give general information about current commercial materials and those currently under development. Detailing their properties, processing methods and how they may affect the future of dentistry.

Keywords: Glass-ceramics, properties, commercial ceramics, dentistry, processes, glass-ceramics for dental restorations, ideal properties for dental glass-ceramic materials, processing and production methods of commercial glass-ceramic materials.

1. INTRODUCTION

The focus of dentistry at present is not only on the prevention and treatment of disease, but also in meeting the demand for better aesthetics. There is currently a significant interest in all-ceramic dental restorations because of the public desire to replace metals as the primary load-bearing tooth restorative material and to improve aesthetic appearance. Traditional metal-based restorations are increasingly being shown to be unsafe or are perceived as unsafe. For example,

*Address correspondence to Anthony Johnson: Academic Unit of Restorative Dentistry, School of Clinical Dentistry, University of Sheffield, Claremont Crescent, Sheffield S10 2TA, UK; Tel: +44(0)114 271 7940; Fax:+44 (0)114 2265484. E-mail: a.johnson@sheffield.ac.uk

Sooraj H. Nandyala and José D. Santos (Eds)
All rights reserved-© 2013 Bentham Science Publishers

mercury containing amalgam restorations may cause some problems by the release of this metal and the nickel chromium alloy used for porcelain fused to metal restorations can be dangerous as it can cause allergic contact dermatitis [1]. All-ceramic restorations are attractive to the patient and the dentist because of their excellent aesthetic, which is similar to the natural tooth. Besides their appearance, metal-free materials have good colour stability, high strength, high chemical resistance and biocompatibility [2]. Currently a number of all-ceramic systems are available to be used as dental restorations.

The first ceramic tooth material used to produce denture teeth was patented in 1789 by the French dentist de Chemant [3]. This product was in fact an improved version of "mineral paste teeth" that was produced in 1774 by Duchateau. It was introduced in England soon thereafter by de Chemant. This baked compound was not used to produce individual teeth at the time, because there was no effective way to attach the teeth to the denture base material. The first ceramic crown arrived in 1903 and was made by Dr Charles Land [4]. These crowns were made using a technique employing a platinum foil matrix and high–fusing feldspathic porcelain. These crowns exhibited an excellent aesthetic, but the low flexural strength of porcelain resulted in a high incidence of failure [4]. Unfortunately, feldspathic porcelains are too weak to be used reliably in the construction of all-ceramic crowns (those without a cast-metal core or metal-foil coping). Furthermore, their shrinkage during firing causes significant discrepancies in the fit and adaptation of margins of the restorations, unless correcting techniques are used [3]. Recent developments have seen an increase in the use of glass-ceramic systems, which are becoming one of the most popular systems because of their superior properties and good marginal fit, compared to traditional feldspathic porcelains [5].

In the 1950's, the addition of leucite to porcelain increased its coefficient of thermal expansion (CTE) and enabled their fusion to gold alloys substructures [6]. The first commercial dental ceramic was developed by Vita Zahnfabrik in 1963 [3]. Although the first Vita ceramic products were known for their aesthetic properties, they still had limited use with different alloys, because of their incompatible CTE. Since then, developments in dental ceramics such as opalescence, specialized internal staining techniques, greening-resistant ceramics and ceramic shoulder margins have significantly enhanced the overall appearance of dental ceramic crowns and bridges, along with the clinical longevity of these prostheses.

The first alumina-reinforced core, consisting of a glass matrix containing 40 to 50 wt% aluminium oxide, was introduced by McLean and Hughes in 1965 [7]. The core was baked on a platinum foil and subsequently veneered with expansion-matched porcelain. The alumina particles improved the fracture resistance of the material, making it stronger, more effective at preventing crack propagation and acted as crack stoppers [8]. Unfortunately, because the core material was opaque and chalky white in appearance, veneering with feldspathic porcelain was necessary for aesthetic reasons. Even though the flexural strength was increased to 120-150MPa, compared to only 60MPa of feldspathic alone, it was still only recommended for the anterior region of the mouth [8]. In 1983, Dr Horn proposed the use of hydrofluoric acid, as an etchant for veneers constructed from a leucite-containing feldspar, to enhance the bond between the ceramic and the resin based composite [9].

Improvement in all ceramic systems by controlled crystallization of the glass (Dicor) was demonstrated by Adair and Grossman in 1984 [10]. The new CAD/CAM technology, by scanning dies without the need for a wax pattern, was introduced in mid 1980s and is still evolving [11]. In the early 1990s, a pressable glass-ceramic (IPS Empress) containing approximately 34 vol% leucite was introduced that provided the strength and marginal adaptation similar to those of the Dicor glass-ceramic, but required no specialized crystalline treatment [12]. Subsequently, a more fracture resistant version called IPS Empress II containing 70% volume of lithium disilicate glass-ceramic was introduced in the late 1990s. Since then, glass-ceramics have become a fundamental material for dental restorations, especially where optimal aesthetics are required.

Glass-ceramic materials are polycrystalline solids containing a residual glass phase, prepared by melting glass and forming it into products that are subjected to controlled crystallization. The concept of controlled crystallization of glass designates the separation of a crystalline phase from the glassy parent phase in the form of tiny crystals, where the number of crystals, their growth rate and final size are controlled by suitable heat treatment. The initial glasses for the preparation of the glass-ceramic materials usually start from inorganic raw materials such as silicon dioxide, potassium carbonate and magnesium silicate [13]. Therefore, the glass-ceramic may demonstrate special morphologies related to their particular structures, as well as considerable differences in appearance depending on their mode of growth. All these different ways of forming microstructures involve controlled nucleation and crystallization, as well as the choice of parent glass composition [14]. However, to ensure a high strength for the glass-ceramic, it is important that the crystals are numerous and are uniformly distributed throughout the glassy phase. Glass-ceramics in dentistry are available as castable, machinable, pressable and infiltrated forms, which are used in all-ceramic restorations. These materials have excellent mechanical and aesthetic properties for the fabrication single crowns, inlays and onlays and three-unit bridges.

2. GLASS-CERAMICS FOR DENTAL RESTORATIONS

Glass-ceramics are can be used as biomaterials in two different areas; firstly, they are used in restorative dentistry as highly durable materials for dental restorations and secondly, they are used for the replacement of hard tissue as bioactive materials [15, 16].

The biomaterials that are used in restorative dentistry have to be durable in the oral environment, exhibit high strength, wear resistance and have the appearance of the natural tooth structure [15, 16]. Restorative materials in dentistry are used to replace the enamel and/or dentine portion of the tooth and to reconstruct the shape and function of the tooth. Nowadays, the focus of dentistry is not only on the prevention and treatment of disease, but also on meeting the demands for better aesthetics. The increasing demand for aesthetic materials in dentistry has led the development of novel all-ceramic systems. Glass-ceramics are one of the materials that are used to produce metal free restorations because of their good aesthetic, marginal fit, mechanical and chemical properties and lack of porosity [17].

The glass-ceramic systems were first discovered in the mid 1950s by Stooky [18], who was attempting to achieve a permanent photographic image by precipitating silver particles in lithium silicate glasses. Their glass transition temperature was approximately 450°C, but when the furnace was accidentally overheated to 850°C a white material was produced, which had a high strength and was called a glass-ceramic. The principal introduction of glass-ceramics into dentistry was in 1968 by MacCulloch [19], who constructed posterior denture teeth, which was based on a Li_2O. ZnO. SiO_2 system.

The use of leucite-based sintering ceramics for veneering frameworks was launched in 1984. Known as Dicor, it was developed by Corning Glass Works, Dentsply, International, PA, USA from a formulation of a low thermal expansion ceramic used for cookware [20]. This glass-ceramic was based on the tetrasilicic mica composition $K-x$ $Mg_{2.5} + x/2$ $Si_4O_{18}F_2$, where x is less than 0.2 [20]. The material contained 45% glass (in volume) and 55% crystalline tetrasilicic fluormica. Some commercially available fluormica glass ceramics are shown in Table **1**.

Table 1: Commercial fluormica glass-ceramics

Ingredients	Macor Corning	Dicor Corning	Bioverit I	Bioveri II
SiO_2	47.2	60.9	38.7	44.5
B_2O_3	8.5	--------	--------	-------
Al_2O_3	16.7	0.6	1.4	29.9
MgO	14.5	17.1	27.7	11.8
K_2O	9.5	13.8	6.8	4.9
F	6.3	4.9	4.9	4.2
Zro_2	--------	4.7	--------	-------
CeO_2	---	0.05	--------	-------
Na_2O	--------	-------	--------	4.4
CaO	--------	-------	10.4	0.1
P_2O_5	--------	-------	8.2	0.1
TiO_2	--------	-------	1.9	-------

Dicor is translucent and is claimed to have similar wear properties and hardness to natural tooth enamel. By having a similar composition to enamel and dentine, the materials should also have the capacity to be chemically bonded to tooth material. Table **2** given by McLean 1979 [17] shows hardness and strength compositions for Dicor, dentine and enamel.

Table 2: Mechanical properties of dental enamel, dentine and Dicor dental material

Material	Micro-Hardness	Modulus of Rupture	Compressive Strength
Dicor	360 (k100)	150 MPa	380 MPa
Tooth Enamel	340	10	400
Dentine	70	50	300

This material was processed by a combination of conventional lost wax casting technology and glass casting. It was used to produce veneers, inlays and crowns in the anterior region only, as it did not have sufficient strength for posterior restorations [21].

The material demonstrated good strength, colour stability, resistance to crack propagation, thermal shock and abrasion resistance, as well as excellent biocompatibility [21]. Later, a machinable form of this material, Dicor machinable glass ceramic (MGC) with improved properties was produced, which contained 70% in volume of tetrasilicic fluormica and was supplied in a form of blocks or ingots to be used with a CAD-CAM milling machine [3]. However, because of the low tensile strength and the need to colour the Dicor restorations using an extrinsic stain, some limitations have been encountered with Dicor materials. In addition, the fit of Dicor restorations was deemed to be substandard when compared to metal-ceramic restorations, even if they were still below the 100μm recommended limit [22].

3. CONSIDERATIONS AND IDEAL PROPERTIES FOR DENTAL GLASS-CERAMIC MATERIALS

Bones and teeth, the hard tissues in the human body, have an inorganic component that primarily consists of hydroxyapatite ($Ca_{10}(PO_4)_6(OH)_2$, HA). Enamel, the outer layer of teeth, is the hardest material in the body and therefore it is not surprising that it consists of approximately 92% hydroxyapatite. Teeth function occurs in one of the most inhospitable environments in the human body. They are subject to larger temperature variations than most other parts, coping with the cold of ice (0°C) through to hot drinks and soups. They also encounter pH changes in the range of 0.5 to 8 and encounter stresses associated with chewing, where cyclic stresses may vary from 20 to about 100MPa. Materials that are to be used in the mouth, and need to survive in the environment described above, have to meet many requirements. The material should have good mechanical properties, such as high strength and fracture toughness [3].

The mechanical properties of glass-ceramic materials are believed to be highly effected by the [8, 13, 14]:

- Particle size of the crystalline phase.

- Volume fraction of the crystalline phase.

- Interfacial bond strength between phases.

- Differences in elastic modulus.

- Differences in thermal expansion.

One of the important mechanical properties is that of hardness. The material used to mimic natural tooth dentin tends to be much harder than natural dentin. As a result, the restoration can cause undue wear on the natural teeth. It is, therefore, essential that the restoration should be as smooth as possible to minimize wear.

The use of glass-ceramic materials to produce dental restorations is moving away from glass-ceramic coatings on dental alloys (ceramic fused to metal) towards glass-ceramic on ceramic bodies (full ceramic restoration). The principal reason for this is aesthetic related. It is, therefore, essential that the colour of the restorations be as similar as possible to that of natural teeth [23]. Moreover, the glass-ceramic material should be biologically compatible with the oral environment. As well as being safe, for both the patient and dentist, the material should not be susceptible to contact damage, as this could result in crack formation. The material should not dissolve, erode or corrode and should have physical properties similar to natural enamel and dentine. In addition, the material needs to be produced economically to the required dimensional tolerance [24].

It has been suggested that dental glass-ceramic materials should have the following characteristics [3, 19, 24, 25]:

- A simple processing technique.

- Precise reproduction of the wax model and high stability of form during further firings (staining, glazing *etc.*).

- Adequate strength, including a built in safety factor to withstand functional loading.

- Minimal shrinkage.

- Chemical stability under intraoral conditions.

- Optimal bonding with all current clinically proven alloys (whether they be precious, palladium-silver, palladium-copper or non- precious).

- Natural looking, with vital colouring under all lighting conditions through natural fluorescence and perfect colour stability, even after many firings.

- Have the ability to be reseated after initial shaping.

- A fine texture that presents no problems for grinding and polishing.

For clinical success, the material used for dental glass-ceramic restoration must meet the following requirements [8, 17, 26, 27]:

- High flexural strength.

- High fracture toughness.

- Homogeneity of the microstructure (absence of microcracks, pores, *etc.*).

- Flawless processing, formable.

- Harmless ingredients (mainly oxides of silicon, aluminium, sodium and potassium, *etc.*).

- Very low solubility.

- High stability in the oral environment, high resistance to acidic foods and solutions.

- Low tendency to plaque formation.

- No undesired interaction with other dental materials.

- No chemical decomposition involving the release of a decomposition product.

Dental glass-ceramic materials have a good reputation as a biocompatible material. There have been numerous improvements in the reliability of all-ceramic restorations, and it is estimated that improvements will continue causing this type of restoration to increase its proportion of the restorative material market.

4. CURRENT GLASS-CERAMIC SYSTEMS AND MATERIALS USED IN DENTISTRY

There are a wide range of ceramic materials and processing routes used in dentistry for the construction of restorations. However, a sensible way to classify dental ceramics according to glass phase is as follows [2, 28]:

1. Dental ceramic containing glass ceramic.

2. Dental ceramic containing glass-infiltrated ceramic.

3. Dental ceramic containing a high strength ceramic.

4.1. Dental Ceramic Containing Glass-Ceramic

Dental ceramic containing glass-ceramic is dental ceramic which consists of glass matrix composites derived from natural or synthetic sources surrounded by at least one type of crystal *e.g.* leucite, fluoroapatite or lithium disilicate [2, 3]. These ceramics are used as veneers, cores and body ceramics. Commercially available systems of veneering ceramics include VM7, VM9, VM13, VM15 and VITAPM 9 (Vita Zahnfabrik, Germany), IPS n DESIGN, IPS inline, IPS classic and IPS Empress system (Ivoclar, Liechtenstein), Duceram plus (DeguDent, USA), Noritake (Noritake Dental supply, Japan). The processing of veneering ceramics involves mixing the dental ceramic powder materials with modelling liquid to form a paste, which is then built up to the required shape [17]. Other glass-ceramics in this category *e.g.* e.max CAD and e.max Press are used for veneers, inlays, onlays, anterior single crown units, all-ceramic three unit anterior bridge and premolar bridges. Different processing methods are used to construct these restorations, such as hot pressing and CAD/CAM.

4.2. Dental Ceramic Containing a Glass-Infiltrated Ceramic

Different oxides reinforce the glassy matrix to create a dense glass-infiltrated ceramic [29]. The commercially available examples of these systems are:

1. In-ceram Alumina (Al_2O_3) (Vita Zahnfabrik, Germany).

2. In-ceram Spinel (Vita Zahnfabrik, Germany), which is infiltrated with spinel oxide ($MgAl_2O_4$) and is slightly weaker than In-ceram Alumina, but has improved optical properties.

3. In-ceram Zirconia(Vita Zahnfabrik, Germany), in which one third of the aluminium oxide is replaced by zirconia oxide and it is significantly stronger than In-ceram Alumina [2, 30]. Computer-aided milling and subsequent infiltration of industrially prefabricated blanks are used for processing. This group of dental ceramics is suitable for the manufacture of anterior and posterior single crowns.

4.3. Dental Ceramic Containing Oxide Ceramic or High Strength Ceramic

Usually have pure alumina oxide (Al_2O_3) or zirconium oxide (ZrO_2) formed in the glass matrix [31]. A number of zirconium oxide systems have appeared on the market in recent years, such as Cercon (DeguDent, Germany), Procera All Zirkon (Nobel Biocare, Sweden), DCS (DCS Dental AG, Switzerland) and Lava (3M ESPE, Germany). The material available for densely-sintered high alumina oxide ceramic is Procera All Ceram (Nobel Biocare, Sweden). Computer aided milling of densely sintered blanks is used in the alumina processing, and pre-sintered or densely sintered blanks in the zirconia processing. Alumina can be used as cores for anterior and posterior single crowns, while zirconia can be used as core for single both anterior and posterior crown, as well as bridge frameworks [32].

5. PROCESSING AND PRODUCTION METHODS OF COMMERCIAL GLASS-CERAMIC MATERIALS

A variety of different processing methods have been used over the years to fabricate glass-ceramic restorations. These include:

1. The lost wax casting technique.

2. Layered sintering, developed to produce cast cores in metals, which are subsequently veneered with dental porcelain material (feldspathic materials).

3. Hot pressing is a processing method that was developed by Ivoclar Vivadent (Liechtenstein) and was first described by Wohlwens and Scharer in 1999 [33]. It has been used for industrial ceramics and both dental porcelain fused to metal and all-ceramic restorations [34]. This technique was first introduced for the construction of single unit crowns, inlays, onlays and veneers, using pre-cerammed and pre-coloured glass-ceramic ingots.

4. More recent approaches to the production of dental restorations is the application of computer-aided design/computer-assisted manufacture (CAD/CAM) technology, which represents a major breakthrough for both dental laboratories and surgeries [35].

5.1. The Lost Wax Casting Technique

The lost wax casting technique has been used to produce cast items in metals since the ancient times. Dental castings have to be cast into a predetermined shape dictated by the type and extent of tooth preparation. This, in turn, depends on the extent of material removed to treat carious disease or traumatic damage. The casting method survived unaltered into the twentieth century, apart from improved furnace technology fuel [36] and dental investment materials. Further development in the casting procedure is required to deal with shape complexity and to overcome the high surface tension of gold to produce a dental restoration by casting. Casting machines were introduced which used nitrous oxide to generate the pressure needed to force molten material into the mould [37]. At the same time, development of a centrifugal casting machine by Jameson in 1907 made use of centrifugal force to caste restorations. The first vacuum casting machine was introduced by Dr. Frink in 1908, followed by Ransom and Randolph's in 1920 [38]. It has been reported that this is the most popular method of casting in the UK [39].

The process requires the production of a mould from a carved wax pattern that includes sprues and air risers for the escape of gases. The wax pattern is invested using an appropriate material. The glass-ceramic frit material is put in an alumina or platinum crucible and is melted at a specific temperature according to the type of glass-ceramic material that needs to be cast. This is then cast into the mould using the centrifugal casting machine.

5.2. The Sintering Process

The sintering process is defined as the transformation of an originally porous compact to a strong, dense ceramic [40]. Sintering happens at a temperature above the softening point of the porcelain, where the glassy matrix partially melts and the powder particles coalesce. The outer molecules of these particles become active and their surfaces begin to bond at the point of contact. This migration of molecules leads to growth of the glass areas, movement of grain boundaries and a reduction in porosity. During sintering, the shift in grain boundaries results in the formation of a closely interlocking crystalline structure of considerable strength. A slurry of the ceramic powder is applied to a refractory die or platinum foil, dried and subsequently fired in a porcelain furnace. During sintering, the density of the porcelain greatly increases and is associated with a volume shrinkage of 30-40%. Porosity can be reduced from 5.6 to 0.56% by vacuum firing [30]. However the remaining micropores can create problems, such as cracking or chipping, in the dental restoration. Multiple layering techniques can be used to correct shape, colour and translucency of the restorations. An example of a sintered leucite reinforced glass-ceramic is Fortress (Mirage, Kansas, USA) [3].

5.3. The Hot Pressing

The hot pressing of a ceramic material to form a dental restoration was first described by Wohlwens and Scharer in 1990 [41]. This technique was first introduced for the construction of single unit crowns, inlays, onlays and veneers, using pre-cerammed and pre-coloured glass-ceramic ingots and was developed by Ivoclar Vivadent (Liechtenstein) [41]. To produce a mould into which the leucite reinforced glass-ceramic ingot is pressed, the lost wax casting technique is used where an accurate wax pattern of the restoration is produced and then invested in a refractory die material [40]. The wax pattern is then burnt out to create the mould to be filled by the glass-ceramic. As the glass-ceramic comprises a certain volume of glass phase, the material can be pressed into a mould using the law of viscous flow [16]. The material ingots are pressed with an alumina plugger at 1150°C under a pressure of 0.3- 0.4MPa into the created refractory mould and held for 20 minutes [42]. By the addition of a second crystal phase within the material, the mechanical properties of the material may be improved by reduced crack propagation. In addition, hot pressing helps avoid large pores caused by non-uniform mixing and avoids extensive grain growth or secondary crystallization.

The hot pressed systems have reported flexural strengths of 122.8 MPa for Empress I shaded ceramics [43] and 138 MPa for Optimal heat pressed ceramics [44]. The IPS Empress system (Ivoclar Vivadent AG, Schaan Liechtenstein), first utilized custom made leucite containing ceramic ingots for a hot pressing technique and this has been followed by the Optimal Pressable Ceramic system (Pentron, USA). Pressable glass ceramics became popular because of their good properties including increased flexural strength, decreased porosity, ease of fabrication, occlusal accuracy, excellent marginal fit and their translucency [42, 45].

5.4. The Application of Computer-Aided Design/ Computer-Assisted Manufacture (CAD/CAM) Technology

Is an exciting new development that has taken place over the last two decades. Many methods have been used to collect 3D data of the prepared tooth. One of these methods of fabrication is the machining of restorations using milling CAD-CAM machines. The term CAD/CAM comes from machine-tool technology and stands for "Computer-Aided-Design/Computer-Aided-Manufacturing". This technique re-presents a major breakthrough for both dental laboratories and surgeries. There are several advanced dental CAD-CAM systems available. These include systems, which are capable of direct scanning within the mouth. CAD-CAM technology was first introduced into the field of dentistry in 1971 by Francois Duret, who came up with the idea that the technologies established in industry could be easily transferred to dentistry [46]. The first commercial system, became available in 1980, was the Cerec 1 system introduced by Mormann and Brandestini for use in dental surgeries at the University of Zurich [47]. Subsequently there have been several developments to the system, including the introduction of the second generation in 1994 by Siemens (Bensheim, Germany) [48].

CEREC stands for Ceramic Reconstruction. Continual development of the hardware and software has resulted in rapid improvement in the CEREC system. The Cerec 3 was introduced in 2000. This system operates on a Windows NT or 2000 based platform and is an improved version of the Cerec 2; including the intraoral 3D scanning camera, image processing, computing power and a form-grinding unit. The Cerec 3 can fabricate inlays, onlays, crowns and veneers. When producing the restorations, the new three-dimensional software allows for much easier handling, interpretation and manipulation [49].

The system takes pictures of the prepared cavity or tooth by using a miniature video camera and the computer produces a three-dimensional image. Then by using the software programme, the restoration is designed and milled from a ceramic block placed in the milling unit of the Cerec system. It was the first chair-side CAD/CAM system available [50].

The Cerec 3 system offers chair-side production and thus, single visit placement of indirect restorations without the need for taking impressions, temporaries or any dental laboratory work [47]. Restorations are machined from commercially produced ceramic blocks (Fig. **1**), which are prefabricated under optimum and controlled conditions for producing posterior restorations only [51].

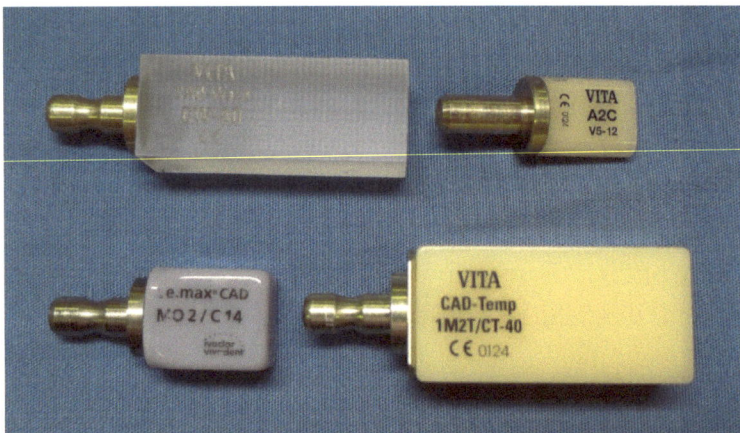

Figure 1: A variety of Cerec blocks are shown. Top right is a material for producing crown structures, which can then be cast into metal using the lost wax casting method; top left is a material for producing veneers; bottom right is a material for producing core structures, which can then be veneered with conventional dental porcelain and bottom left is a material for producing long term temporary bridges.

Cerec blocks are a high and uniform quality ceramic material, without the defects associated with manually produced restorations [52].

The advantages of the Cerec blocks include:

- Consistent material quality due to a standardized manufacturing process.

- Very easy to grind, facilitating contouring and finishing.

- Appearance similar to natural tooth structure.

- Natural translucency.

- Wear resistant and durable in the oral environment.

- The blocks are homogenous and without porosity and can be easy polished to a high luster.

However, there are many disadvantages associated with CAD/CAM systems, including the sensitivity of the technique in both scanning and designing the restorations, which entails a high degree of training, along with the high cost of the equipment.

A number of laboratory based CAD/CAM systems have become available recently as a result of the intensely improved performance of computer hardware and software to produce a restoration by the indirect approach (Cerec InEos scanner and Cerec inLab, Sirona). The Cerec inLab, first introduced in 2002, is designed to produce multiple-unit restorations using high strength ceramic frameworks for crowns and bridges, as well as inlays and crowns. The separate milling unit is connected to the optical unit by radio control. The milling unit receives data from the control unit, according on its location in the office. From the impression, a stone model is poured using laser-visible stone and then scanned automatically with the laser scanner of the inLab system or by using the separate InEos scanner. A three-dimensional image of the model is then produced and the substructure or restoration can then be designed. The software allows customization, such as anatomically shaped frameworks and occlusal surface design based on biogeneric tooth models [53]. After finishing of the design, the selected material block is put into the milling chamber and automatically machined. In the case of frameworks, any necessary sintering is undertaken before an appropriate veneering ceramic is applied.

InEos is an advanced 3D scanner used in conjunction with a dedicated computer and inLab 3D software. The scanner can be used as stand-alone unit or in combination with inLab milling unit to complete restorative work production, while the previous systems could only undertake scanning or milling at one time. This scanner has an automatic image capture function, which allows free movement of the dental model in any direction, providing complete control of the angle of the scan. Thus, the user can take a digital impression of the required treatment area only, which leads to substantial time saving. In addition, the data can also be saved and export scans in an STL file format so it can be used with various brands of software [54].

5.5. E4D

Laser technology: The E4D system is an easy to use chairside CAD/CAM system that is empowering modern dentistry. This technique allows the fabrication of metal-free all porcelain crowns, inlays, onlays and veneers in the same day. With this technique, no impressions are required and no temporaries or second visits are needed. The restorations produced look and feel natural.

After preparing the tooth for a new restoration, the laser scanner takes images of the preparation to create a virtual model in the E4D system. A restoration of the exact specifications of the prepared tooth model fit can be created using a variety of tools in the design software. The information is then sent wirelessly to the E4D milling machine, where diamond tipped milling tools shape the new tooth from a solid block of porcelain that has been matched to the shade of the natural tooth. The milling times are between 10 and 20 minutes depending on the size of restoration being fabricated. Then the new restoration is adjusted occlusally, polished, glazed and bonded to the tooth with dental resin adhesives [55].

6. COMMERCIAL GLASS-CERAMIC RESTORATIVE MATERIALS

IPS Empress (Ivoclar Vivadent) is a leucite-reinforced hot-pressed glass-ceramic that was released onto the market in the early 1990s [33]. The chemical composition is based on the formula of SiO_2–Al_2O_3–K_2O and the microstructure consists of evenly dispersed leucite (KAlSi2O6) crystals embedded in a glassy matrix. This leucite reinforced glass-ceramic obtains its strength by finely dispersed leucite crystal reinforcement and was produced for restoring one unit including veneers, inlays, onlays and crowns [45]. The leucite content ranges from 38 to 40% (in volume) in the glassy matrix [40]. The final restorations may be completed by the application of stains and glazes, or alternatively trimmed back and veneered with thermally compatible ceramics prior to glazing [44]. Recently, the leucite-based material has also been recommended by the manufacturer to be used in the layering technique. The restorations are subsequently bonded to the tooth structure with a luting material, preferably an adhesive bonding system. After additional firings, the flexural strength of Empress significantly improved [56]. The strength increase is attributed to a good distribution of the fine leucite crystals, as well as the compressive stresses arising from the thermal contraction mismatch between the leucite crystals and the glassy matrix [40].

The mechanical properties and microstructure of Empress glass-ceramics processed according to manufacturer's recommendations are well documented in the literature. However, because it was found that this material does not offer the improved strength that the manufacturers claimed and is prone to fracture in posterior regions, there has been a decline in usage of Empress [57]. Moreover, few studies have evaluated glass-ceramics processed under different hot-pressing temperatures. It has been demonstrated that the microstructure, and consequently the mechanical properties of a glass-ceramic can be modified by varying the heat-treatment to which it is submitted [14]. For instance, it has been shown that the final crystal growth took place during the pressing and firing steps of the processing method [56].

Lithium disilicate is a second-generation glass-ceramic and the main representative of this category is the Empress II core material (Ivoclar Vivadent, Schaan, Liechtenstein). Like Empress, this glass ceramic is pre-cerammed by the manufacturer and supplied in ingots for pressing in a furnace. This material contains lithium disilicate ($Li_2Si_2O_5$) as the main crystalline phase, that makes up about 70% of the volume the microstructure of glass ceramic [32]. The microstructure of this material is unusual because it consists of many small interlocking plate-like crystals that are randomly orientated. The interlocking nature of the crystals, as well as their high density, gives this glass ceramic very high flexural strength. The flexural strength ranges between 350 and 450 MPa and fracture toughness between 2.8 and 3.5 MPa/ $m^{1/2}$. This makes the core strong enough to produce crowns for molars and adequate for the fabrication of anterior three unit bridges. The framework of this material can be made-up, either with the lost-wax casting technique using the centrifugal casting machine, heat-pressure technique, or can be milled out of prefabricated blanks using the CAD-CAM milling machine. To enhance the strength and longevity of this material, it is advised that the restorations should be etched and adhesively luted [58].

To improve the aesthetic and wear properties of this material, an apatite-containing glass-ceramic veneer can be sintering to the core. The more translucent Empress II core material eliminates the use of opaque alumina or a metal substructure and potentially gives good aesthetics [44]. The material is indicated not only for the fabrication of anterior bridges, but also for short-span posterior bridges (pontics not wider than a premolar) extending up to the second premolar. Esquivel-Upshaw *et al.* [59] reported a survival rate of 93% for posterior Empress II FPDs after 2 years. Marquardt and Strub [60], reported a survival rate of 100% for single crowns and 70% for FPDs extending up to the second premolar after 5 years of function [61].

Vitabloc Mark II (Vita Zahnfabrik) blocks are made from fine structure feldspar ceramics. Introduced in 1991 as the second generation of ceramic blocks, it replaced the original Vitabloc Mark I ceramic block. The fine particle size microstructure creates a nearly pore-free ceramic, which improves the mechanical properties, polishability and decreases enamel wear. It is considered to be one of the least abrasive dental ceramics. The flexural strength is twice as strong as that of conventional feldspar porcelains, which is about 160 MPa after glazing. This material is suitable for veneers, anterior single unit crowns, inlays and onlays. These blocks can be machined with both the Cerec 3 and inLab systems [32].

ProCAD (professional computer assisted design), manufactured by Ivoclar Vivadent, is a CAD/CAM leucite reinforced glass-ceramic and has a composition of SiO_2-Al_2O_3-B_2O_3-Bao-CaO-CeO_2-$K2O$-Na_2O-TiO_2 and pigments. This material has a flexural strength of 140 MPa and a fracture toughness of 1.3 MPa. $m^{1/2}$. In addition, the chemical solubility of this material is low at $<100\mu g/cm^2$. It has a thermal expansion coefficient of $17.00\mu m/m.k$. After machining the surface of the glass-ceramic can be improved by polishing or applying a specially developed ProCAD glaze. The flexural strength is improved to 180-200 MPa [62]. The glass-ceramic could be etched by applying hydrofluoric acid and to form a retentive layer for adhesive cementation. ProCAD is available in a five different shades and translucencies and characterisation can be achieved with external stains. To achieve adequate stability, glazing the crowns is required. This material could be used in the form of blocks to produce anterior and posterior crowns, inlays and onlays using the Cerec$^®$ 3 and inLab system. It has good stability, a good aesthetic and accurate machinability, because of the small uniform size of the crystals [63].

7. THE FUTURE OF DENTAL GLASS-CERAMIC

The standard of aesthetics in dental glass-ceramics has now reached a stage where the development of new glass-ceramics is needed to meet ever-increasing demands on their optical and physical properties. More recent advancements in dental materials and improvements in fabrication of resin-bonded glass-ceramic restorations provide real opportunities for achieving excellent aesthetics. However, these indirect restorations also have limitations. Possible signs of failure include de-bonding and fracture of the material, particularly in relation to cementation procedure [10-12]. In addition, clinical longevity remains a concern. Especially in the posterior region, where insufficient strength and fracture resistance are major concerns [64]. Ceramics are inherently brittle materials and are susceptible to catastrophic failure. Therefore, a future aim for dental glass ceramics should be that of strengthening the glass ceramic without sacrificing aesthetics. Recently, a number of novel dental glass-ceramics are under development.

7.1. Apatite-Mullite Glass-Ceramics

Apatite-mullite glass-ceramics are glass ionomer cement derivative glasses first produced by Hill *et al.* (1991) [65] at The University of Limerick and have since been developed in many universities around the world. This glass-ceramic based on the formula $4.5SiO_2-0.5P_2O_5-A_{12}O_3-xCaO-xCaF_2$, where x varied from 0 to3. These glasses, when subjected to a heat treatment, were found to crystallize to an apatite phase (Fluorapatite- $Ca_{10}(PO_4)_6F_2$) and a mullite phase ($Al_6Si_2O_{13}$). They were found to undergo glass in glass phase separation, known as amorphous phase separation. The high fracture toughness is thought to arise from the microstructure, which consists of interlocking apatite and mullite crystals [66].

Essentially, it has been shown that apatite-mullite glass-ceramics can be used to produce a restoration in one working day, it can be cast, cerammed while still in the casting mould and then de-vested and trimmed within a working day. Some studies showed that this material has good mechanical and chemical properties and could be processed to produce a restoration using the lost wax casting technique and the CAD-CAM system with good marginal fit [67, 68].

7.2. Fluorcanasite Glass-Ceramics

Fluorcanasite glass-ceramics have been investigated by van Noort *et al.* [69] and have illustrated that canasite, derived from the stoichiometric composition has potential for the fabrication of dental restorations. These fluorcanasite glass ceramics were found not to be suitable for the lost wax technique and press system, but were successfully milled using CAD/CAM technology. These glass ceramics show a microstructure of bulk crystal phases and high flexural strength and fracture toughness with outstanding aesthetic properties.

7.3. The Latest Development Leucite Glass Ceramic

The latest development leucite glass ceramic is known as a strong glass ceramic [70]. This dental glass ceramic contains small, uniformly dispersed, single leucite crystals of ellipsoidal habit and very uniform particle size. When used as a powder, the glass ceramic can be used with the platinum foil or refractory investment technique to produce dental restorations or it can be pressed and sintered into blocks or ingots. It can be used with either the lost wax casting technique or with CAD/CAM techniques to produce very strong and aesthetic restorations. The strength of this material is illustrated by a high flexural strength of 245 MPa compared to a flexural strength of 166 MPa for the widely used commercial material Empress Esthetic.

8. CONCLUSION

Glass-ceramics for dental applications show many advantages over dental all ceramic materials. Dental glass ceramics have now reached a stage where the development of new materials is being actively pursued to try to meets the ever-increasing demands on their aesthetic and physical properties. In addition, the restorations can be processed in many different systems such as sintering systems, 3D printing systems, pressing and CAD/CAM technology. However, some clinical restoration failures have been reported [71] and there are limitations of using all ceramic in posterior teeth, in the main because of fracture toughness and strength issues. Therefore, in the future, glass ceramic research for dental applications should focus on trying to strengthen dental glass ceramics without sacrificing aesthetics and be supplemented by a variety of different machining techniques and production routes.

ACKNOWLEDGEMENT

Declared none.

CONFLICT OF INTEREST

There is no conflict of interest with other people or organisations in respect of the present research work.

REFERENCES

[1]   Olivieri G, Brack CH, Müller-Spahn F, Stohelin HB, Heirman M, Renard P, Brockhaus M, Hock C. Mercury induces cell cytotoxity and oxidative stress and increases ß-amyloid secretion and tau phosphorylation in SHSY5Y neuroblastoma cells. J Neurochem 2000, 74: 231-236.

[2]   Kelly JR, Ceramics in restorative and prosthetic dentistry. Annu Rev Mater Sci 1997, 27: 443-468.

[3]   Anusavice K, Phillip's science dental materials. Eleventh ed: St Louis, Missouri Saunders 2003.

[4]   Land C, Porcelain dental Arts. Dental Cosmos 1903, 45: 615-20.

[5]   Bergman MA, The clinical performance of ceramic inlays: a review. Aust Dent J 1999, 44: 157-168.

[6]   Schmid M, Fischer J, Salk M and Strub J., Leucit-Verstakter Glaskeramiken. Schwiez Monatsschr Zahnmed 1992, 102: 1046-53.

[7]   McLean JW and Hughes TH The reinforcement of dental porcelain with ceramic oxides. Br Dent J 1965, 119: 251-267.

[8]   van Noort R, Introduction to Dental materials. 3rd ed. 2002, Mosby.

[9]   Horn HR, Porcelain laminate veneers bonded to etched enamel. Dent clin North Am 1983, 27: 671-684.

[10]   Adair PJ and Grossman D The castable ceramic crown. Int J Perio rest Dent 1984, 4(2), 32-64.

[11]   Kramar N, Frankenberger R, Pelka M, and Petshet A, IPS empress inlays and onlays after four years a clinical study. J Dent 1999, 27(5), 325-331.

[12]   Pallensen U and van Dijken JW, An 8 year evaluation of sintered ceramic and glass ceramic inlays processed by the cerec CAD/CAM system. Eur J Oral Sci 2000 108(3), 239-246.

[13]   Strnad Z Glass-ceramic materials, Amsterdam: ELSEVIER 1986.

[14]   Holand W and Beall G, Glass-ceramic technology, Westerville: The American Ceramic Society 2002.

[15]   Huang CM, Kuo DH, KimYJ and Kriven WM, Phase-stability of chemically derived enstatile (MgSiO3) powders. J Am Ceram Soc 1994, 77: 2625-2631.

[16]   Holand W, Rheinberger V, Apel E, van't Hoen C, holand M, Dommann A, Obrecht M, Mauth C and Graf-Hausner U, Clinical applications of glass-ceramics in dentistry. J Mater Sci 2006 17: 1037-1042.

[17]   McLean JW The Science and art of Dental ceramics; The nature of dental ceramics and their clinical use, Chicago: Quintessence 1979.

[18]   Stookey SD, Catalyzed crystallization of glass in theory and practice. Glastechn Ber 1959, 32: 1-8.

[19]   MacCulloch WT, Advanced in dental ceramics. Brit Dent J 1986, 124: 361-365.

[20]   Hobo S and Iwata T, Castable apatite ceramics as a new biocompatible restorative material : Fabrication of the restoration. Quintessence International 1985, 16: 207-216.

[21]   Grossman DG and Walters HV, The chemical durability of dental ceramics. J Dent Res 1989, 63: 234-574.

[22]   Malament KA and Socransky SS, Survival of Dicor glass-ceramic dental restorations over 14 years. Part II: effect of thickness of Dicor material and design of tooth preparation. J Prosthet Dent 1999, 81: 662-627.

[23]   McLean JW, Dental Ceramic proceedings of the first international symposium on ceramics, Chicago: Quintessence publishing co 1983.

[24]   McLean JW, Dental ceramics proceedings of the first international symposium on ceramics, Chicago: Quintessence 1983.

[25]   Naylor WP, introduction to Metal ceramic technology, Chicago: quitessence 1992.

[26]   Fradeani M, An 11-year clinical evaluation of leucite-reinforced glass-ceramic crown: A retrospective study Quintessence International 2002, 33: 503-510.

[27]   Sadowsky SJ, An overview of treatment considerations for esthetic restorations: A review of the literature. J Prosthet Dent 2006, 96: 433-442.

[28]   Hammerle C, Sailer I, Thomas A, Halg G, Suter A, and Ramel C, Dental ceramics; Essential Aspects for clinical practice, London: Quintessence Publishing 2009.

[29]   Rosenblum MA and Schulman A, A review of All-Ceramic restorations. J Am Dent Assoc 1997, 128: 297-307.

[30]   Kelly JR, Nishimura I,and Campbell SD Ceramic in dentistry : historical roots and current perspectives. J prosthet Dent 1996, 75: 18-32.

[31] Blatz M, Sadan A, and Kern M, Resin-ceramic bonding: a review of the literature. J Prosthet Dent 2003, 89: 268-274.

[32] Conrad HJ, Seong WJ, and, Pesun IJ, Current ceramic materials and systems with clinical recommendations: A systematic review. J Prosthet Dent 2007, 98(5), 389-404.

[33] Wohlwend A, Scharer P, and, Stub JR, Metalloceramic and full ceramic restorations. Die quintessenz 1990, 41: 981-991.

[34] Beham G, IPS Empress: a new ceramic technology. Ivoclar-Vivadent Report 1990, 6: 1-15.

[35] Fasbinder DJ, Clinical performance of chairside CAD/CAM restorations. J Am Dent Soc 2006, 137: 22S-31S.

[36] Hunt LB, The long history of lost wax casting. Gold Bulletin 1980, 13: 63-79.

[37] Taffart WH, A new and accurate method of making gold inlays. Dental Cosmos 1907, 49: 1117-1121.

[38] Hagman HC, The evaluation of metal casting for dentistry. Bulletin Hist Dent 1976, **2:** 98-105.

[39] Johnson A, A survey of casting and melting technique used in the lost wax casting of yellow gold and metal/ceramic alloys in British Dental schools Restorative Dent 1989, 5: 18-23.

[40] Denry IL, Recent Advances in Ceramics for Dentistry. Crit Rev Oral Biol Med 1996, 7(2), 134-143.

[41] Wohlwend A and Scharer P, Die empress-technique. Quintessence zahntech 1990, 16: 966-978.

[42] Gorman CM, McDevit WW, and Hill RG, Comparison of two heat-pressed all-ceramic dental materials. Dent Mater 2000, 16: 389-395.

[43] Zeng K, Oden A, and Rowcliffe D, Flexure tests on ceramics. Int J Prosthodont 1996, 9: 434-439.

[44] Cattell MJ, P.alumbo RP, Knowles JC, Clarke RL and Samarawickrama DY, The effect of veneering and heat treatment on the flexural strength of Empress 2 ceramics. J Dent 2002, 30: 161-169.

[45] Albakry M, Guazzato M, and Swain MV, Biaxial flexural strength, elastic moduli, and x-ray diffraction characterization of three pressable all- ceramic materials. J Prosthet Dent 2003, 89: 374-380.

[46] Miyazaki T, Hotta Y, Kunii J, Kuriyama S, and Tamaki Y, A review of dental CAD/CAM: current status and future perspectives from 20 years of experience. Dent Mater 2009, 28(1), 44-56.

[47] Mormann WH, Brandestini M, and Lutz F, The Cerec system: computer-assisted preparation of direct ceramic inlays in 1 setting. Die quintessenz 1987, **38:** 457-470.

[48] Mehl A and Hickel R, Current state of development and perspectives of machine-based production methods for dental restorations. Int J Comput Dent 1999, 2: 9-35.

[49] Mormann WH, The evolution of the CEREC system. J Am Dent Assoc 2006, 137: 7S-13S.

[50] Strub JR, Rekow ED, and Witkowski S, Computer-aided design and fabrication of dental restorations. J Am Dent Assoc 2006, 137: 1289-1296.

[51] Martin N and Jedynakiewicz NM, Clinical performance of CEREC ceramic inlays: a systematic review. Dent Mater 1999, 15: 54-61.

[52] Jedyankiewicz MN and Martin N, CAD-CAM in restorative dentistry; The cerec Method. 3rd ed, Liverpool: Liverpool University Press, 1993.

[53] Dunn M, Biogeneric and user-friendly: the Cerec 3D software upgrade V3.00. Int J Comput Dent 2007, 10: 109-117.

[54] Kurbad A and Reichel K, InEos-new system component in Cerec 3D. Int J Comput Dent 2005, 8(1), 77-84.

[55] D4D technologies, Scientific documentation E4D laser technology, Texus 2009.

[56] Dong JK, Luthy H, Wohlwend A and Scharer P, Heat-pressed ceramics: technology nad strength. Int J Prosthodont 1992, 5: 9-16.

[57] Holand W, Schweiger M, Frank M and Rheinberger VM, A comparison of the microstructure and properties of the IPS empress and the IPS Empress glass-ceramics. J Biomed Mater Res 2000, 53: 297-303.

[58] Sorensen J, The IPS Empress II system. Definig the possibilities. Quint Dent Technol 1999, 22: 153-163.

[59] Esquivel-Upshaw JF, Anusavice KJ, Young H, Jones J and Gibbs C. Clinical performance of a lithia disilicate-based core ceramic for three-unit posterior FPDs. Int J Prosthodont 2004, 17: 469-475.

[60] Marquardt P and Strub JR. Survival rates of IPS empress 2 all-ceramic crowns and fixed partial dentures: results of a 5 year prospective clinical study. Quintessence Int 2006, 37: 253-259.

[61] Pascal M and Rudolf SJ, Aurvival rates of IPS empress 2 all-ceramic crowns nad fixed partial dentures:results of a 5-year prospective clinical study. Quintessence International 2006, 37(4), 253-259.

[62] Gaglio MA, Esthetic restorations designed with confidence and predictability. Compend Contin Educ Dent 2001, 22: 30-34.

[63] Ivoclar vivadent AG, Scientific documentation ProCAD, Schaan: Liechtenstein 2002.

[64] Ivoclar vivadent AG, The all-ceramics specialist, held a large expert meeting at its headquarters. Dental News 2006.

[65] Hill RG, Patel M, and Wood DJ, Preliminary studies on castable Apatile-Mullite glass-ceramics, Proceedings of the 4th International Symposium on ceramics in medicine, London: Butterworth-Heinmann Ltd, 1991.

[66] Hill RG and Clifford A, Apatite-Mullite glass-ceramics. J Non-Cry Solids 1996, 346-351.

[67] Fathi H, Johnson A, van Noort R, Ward J and Brook I, The effect of calcium florid (CaF2) on the chemical solubility of an apatite-mullite glass-ceramic material. Dent Mater 2005, 21: 551-556.

[68] Fathi H, Johnson A, van Noort R, and Ward J, The influence of calcium florid(CaF2) on Biaxial flexural strength of apatite-mullite glass-ceramic materials. Dent Mater 2005, 21: 846-851.

[69] van Noort R, Shareef MY, Johnson A and James PF, Properties of a canasite glass-ceramic. J Dent Res 1997, 76(21), 61.

[70] Chen X.H, Cattell, M.J., Ibsen R.L., Riddel J.V. Chadwick T.C. Strong glass-ceramic. WO/2009/073079, World intellectual Property Organisation, 11.06.2009.

[71] Boushell LW, Ritter AV, Ceramic inlays: a case presentation and lessons learned from the literature. J Esthet Restor Dent 2009, 21(2), 77-87.

Send Orders of Reprints at bspsaif@emirates.net.ae

CHAPTER 3

Novel Ceramics and Glass-Ceramics by Microwave and Conventional Processing: A Review

Sumana Ghosh*, Kalyan Sundar Pal, Someswar Datta and Debabrata Basu

Bio-Ceramics and Coating Division, CSIR - Central Glass and Ceramic Research Institute, Kolkata-700 032, West Bengal, India

Abstract: Microwave heating technique has attracted considerable attention for the processing of various materials such as ceramics, glasses, polymers, composites and even metals. Researchers are trying to apply this technology to new areas. The present review presents a short overview of some recent applications of conventional and/ or microwave processing for the synthesis of novel ceramics and glass-ceramics.

Keywords: Ceramics, Glass-ceramics, Microwave heating, Conventional heating.

1. INTRODUCTION

Ceramic materials possess unique characteristic features such as high-temperature strength, high hardness, superior wear resistance, lower thermal and electrical conductivity and chemical stability [1]. Glass-ceramics are micro- or nanocrystalline materials prepared by controlled nucleation and crystallization of a glass precursor. The properties of glass-ceramics depend on composition, phase assemblage and microstructure. The phase assemblage is determined by the composition and heat treatment, which in turn governs many properties *e.g.* hardness, density, thermal expansion *etc.* The microstructure also plays an important role in controlling the properties of glass-ceramic materials [2].

Microwaves are electromagnetic waves with wavelengths from 1 mm to 1 m and corresponding frequencies between 300 MHz and 300 GHz. Microwaves are coherent and polarized. They can be transmitted, absorbed, or reflected depending on the type of the material. The microwave energy is being utilized for the processing

*Address correspondence to Sumana Ghosh: CSIR-Central Glass and Ceramic Research Institute, 196, Raja S.C. Mullick Road, Kolkata-700 032, India; Tel: +91(033)-2473 3469/76/77/96, Fax: +91 (033)-2473 0957; E-mail: sumana@cgcri.res.in

Sooraj H. Nandyala and José D. Santos (Eds)
All rights reserved-© 2013 Bentham Science Publishers

of various materials as it offers several advantages over conventional heating methods such as unique microstructure and properties, improved product yield, energy savings, reduction in manufacturing cost and synthesis of new materials [3].

In the conventional heating method, the thermal energy is delivered to the surface of the material by radiant and/or convection heating that is transferred to the bulk of the material *via* conduction. The microwave energy is delivered directly to the material through molecular interaction with the electromagnetic field [3]. The dielectric properties determine the effect of the electromagnetic field on the material [4]. The interaction of microwaves with a dielectric material results in translational motions of free or bound charges and rotation of the dipoles. The resistance of these induced motions due to inertial, elastic, and frictional forces leads to losses resulting in volumetric heating [3]. As microwaves can penetrate the material, heat can be generated throughout the volume of the material resulting in rapid, uniform volumetric heating. Therefore, the thermal gradient in the microwave heated material is the reverse of that in the conventionally processed material. The processing time can be reduced and the product quality can be enhanced by microwave processing. Microwaves can selectively couple with the higher loss tangent material when it is in contact with materials having different dielectric properties. Hence, microwaves can be used for the selective heating of the materials. However, localized thermal runaway may occur during processing of materials in non-uniform electromagnetic fields that leads to high enough stresses and material fracture [4].

Microwave and conventional processing of a variety of conventional as well as advanced materials has been already described [3-5]. The objective of this article is to present some recent applications of microwave and conventional heating method for the processing of novel ceramics and glass-ceramics useful in the near future.

2. PROCESSING OF NOVEL CERAMICS BY MICROWAVE HEATING

2.1. Development of Oxide Coating on Aluminium

Aluminium oxide coating was developed on commercial aluminium using microwave heating technique [6]. Porous, thin coating (~42 μm) and dense, thick coating (~661 μm) was developed by microwave heating of the pre-oxidized (600°C, 200h) aluminium samples for 60 and 90 min, respectively. In both the cases, smooth coating with crack free interface was observed. Microwave heating was also utilized to develop oxide coatings on commercial aluminium samples of various geometrical shapes [7]. Experimental results showed that the volume to surface ratio had a significant effect on microwave induced oxidation behavior of the samples. Coating thickness and density gradually increased with the increase of microwave exposure time. Protective aluminium oxide coating of varied microstructure and thickness was formed on commercial aluminium by suitable adjustments of the microwave processing parameters.

2.2. Synthesis of Al-α-Al$_2$O$_3$ Composite

Though bulk metals act as reflector, finely divided metallic powder absorbs the microwave radiation because multiple scattering coupled with eddy current loss play a significant role in the microwave absorption [8]. Sintering of fine metallic powders, intermetallic compounds, alloys and nanocomposites was achieved by this processing method [8-12].

Al-Al$_2$O$_3$ composites are now widely used for potential applications in the acrospace, defense and automotive industries [13]. Al-Al$_2$O$_3$ composites have been prepared from aluminium and its alloys by various routes such as high temperature (~900-1400°C) oxidation of liquid Al-Mg and Al-Mg-Si alloys [14-16], directed metal oxidation [17], reactive melt infiltration [18], squeeze casting [19, 20], *in situ* method [21], friction stir processing [22], powder metallurgy [23, 24], microwave heating [25] *etc.*

Al-Al$_2$O$_3$ composite was prepared by microwave processing using thermally oxidized Al metal [25]. When this oxidized Al was exposed to microwave radiation, the oxide coating absorbed the microwave energy and became volumetrically heated. Consequently, the metallic layer adjacent to the coating got heated by the conduction method. At higher temperatures, aluminium oxide coating absorbed the microwave energy at an enhanced rate because of higher loss tangent and relative dielectric constant values. Thus, metal is heated to a high temperature during long microwave exposure, which led to activated diffusion resulting in high rate of oxidation. Oxygen atoms diffused through the pores of the oxide coating to the aluminium metal. As a result, aluminium metal was oxidized and formed Al-Al$_2$O$_3$ composite depending on provision for the diffusion of oxygen atoms towards the inner side [25].

Al-α-Al$_2$O$_3$ core-shell composite was synthesized by microwave-assisted oxidation of the commercial Al powder [26, 27]. The as received Al powder and the microwave heated Al powder were characterized by X-ray diffractometry (XRD), fourier transform infrared (FTIR) spectroscopy, scanning electron microscopy (SEM), transmission electron microscopy (TEM), energy dispersive X-ray (EDX) analysis and zeta potential measurement. XRD data confirmed the formation of Al-α-Al$_2$O$_3$ composite. FTIR studies and SEM observations indicated the formation of Al-α-Al$_2$O$_3$ core-shell composite. TEM images, corresponding selected area electron diffraction (SAED) patterns and EDX analysis confirmed the Al-α-Al$_2$O$_3$ core-shell composite formation. Zeta potential measurements also indicated that Al core particles were surrounded with α-Al$_2$O$_3$ shell [26].

Al-Al$_2$O$_3$ composites were developed by microwave processing of undoped and doped Al powder compacts [28]. Mg, Mg-Si and Al$_2$O$_3$ were used as dopants. The microwave treated undoped and doped Al powder compacts were characterized by XRD, SEM and EDX analysis. XRD and EDX data of the microwave processed powder compacts confirmed the formation of Al-Al$_2$O$_3$ composites. Bulk composite growth was observed in the Mg and Mg-Si doped Al powder compacts whereas composite formation was found in the surface only in the case of undoped and Al$_2$O$_3$ doped Al powder compacts. SEM, EDX and TEM of the microwave heated Al powder indicated the formation of Al-Al$_2$O$_3$ core-shell composites in the microwave heated undoped and doped Al powder compacts [28].

2.3. Development of Rutile Titania

Commercial titania powder was sintered by microwave heating to obtain rutile as the major phase [29]. Almost 100% rutile phase was obtained after only 1 h processing at 1300°C in microwave while this conversion was much lower in the conventionally sintered specimens. The maximum density obtained for titania ceramics sintered in microwave for 1 h at 1300°C was ~96% of theoretical density whereas it was ~85% of theoretical density for titania ceramics conventionally sintered under identical conditions in a conventional muffle furnace. Furthermore, microwave processed titania (1300°C, 1 h) showed better dielectric properties compared to the conventionally sintered one. However, DC electrical resistivity was almost comparable for both types of specimens. It was observed that the microwave processed titania ceramics had finer microstructures than those of the conventionally sintered ones. This indicates that microwave processing promotes phase transformation, densification and fine grain size development in the sintered titania ceramics [29].

3. NOVEL GLASS-CERAMICS BY MICROWAVE AND/OR CONVENTIONAL PROCESSING

3.1 Glass-Ceramic Coatings for Gas Turbine Engine Components by Conventional Processing

New high temperature and abrasion resistant glass-ceramic coating based on $MgO-Al_2O_3-TiO_2$ and $ZnO-Al_2O_3-SiO_2$ glass systems was developed for gas turbine engine components [30, 31]. Table **1** shows the compositions of coating materials investigated. The coating materials and the resultant coatings were characterized by differential thermal analysis, differential thermo-gravimetric analysis, XRD, optical microscopy and SEM. Static oxidation resistance at 1000°C for continuous service, thermal shock resistance, adherence at 90°-bend test and abrasion resistance were evaluated using suitable standard methods. XRD analysis of the coating materials and the resultant coatings showed presence of microcrystalline phases. SEM micrographs indicated strong chemical bonding at the interface of metal and ceramic. Optical micrographs showed smooth glossy impervious defect free surface finish. Evaluation of coating properties showed the suitability of these coatings for the protection of different hot zone components in different types of engines [30].

Table 1: Approximate compositions of coating materials studied [30]

Oxide	$MgO-Al_2O_3-TiO_2$ based glass	$ZnO-Al_2O_3-SiO_2$ based glass
SiO_2	30-40	40-50
MgO	10-20	1-3
ZnO	-	30-40
Al_2O_3	15-25	10-20
TiO_2	10-20	-

Table 1: cont....

B$_2$O$_3$	5-15	1-3
CoO	-	0-1
NiO	-	0-1
R$_2$O (R= Na, K)	5-10	-

3.2. Development of Bulk Glass-Ceramics and Glass-Ceramic Coating by Microwave Processing

Monoclinic celsian was obtained from 10% CuO seeded barium aluminosilicate glass by microwave heating at 1300°C for 15 min [32]. Bulk crystallization of glasses belonging to M$_2$O-CaO-SiO$_2$-ZrO$_2$ system (where M=Li, Na, K) was studied using microwave radiation. The microwave heated glass samples showed surface and bulk crystallization while only surface crystallization was observed in the glass samples processed by conventional heating method [33]. The sintering and devitrification in the ternary CaO-ZrO$_2$-SiO$_2$ glass system was also investigated at 900-1050°C by microwave processing [34].

Hard glass-ceramic coating was formed by microwave heating of a glass coating based on MgO-Al$_2$O$_3$-TiO$_2$ system on nickel based superalloy (Nimonic - AE 435) substrate [35-37]. The coating was characterized by XRD, SEM, image analysis, surface roughness measurement and hardness evaluation by depth sensitive indentation (DSI) technique. Surface and cross-sectional SEM examinations of the samples demonstrated that the microwave heated coating contained finer crystallites than the sizes of the crystallites present in the conventionally processed coating under identical processing conditions. The microwave treated glass-ceramic coating showed lower surface roughness (R$_a$) value than that of the conventionally processed glass-ceramic coating. DSI results showed that the microwave processed coating had much higher hardness (~6 GPa) compared to that (~5 GPa) of the conventionally treated coating [35].

3.3 Conventionally Processed Glass-Ceramic Based Bond Coat for Thermal Barrier Coating System

Thermal barrier coating (TBC) is required for the thermal protection of nickel base superalloys used as gas turbine blades [38, 39]. Usually, thermal barrier coating system is a three layered structure consisting of one ceramic top coat, an intermediate NiCoCrAlY/PtAl based metallic bond coat and a metallic substrate. In order to protect the underlying metal from oxidation and high temperature corrosion and to enhance the adherence between the metallic substrate and the ceramic top coat metallic bond coat is deposited between the substrate and the top coat [40].

During service, thermally grown oxide (TGO) scale is produced at the bond coat-top coat interface due to oxidation of the bond coat at the elevated temperatures. The TGO layer thickness increases with increasing the operation time. This phenomenon is the most significant factor in determining the lifetime of a TBC system. High stresses are generated in the bond coat-TGO interface because of oxide thickening (volume increase), thermal expansion misfit and applied loads. Consequently, crack initiates and propagates that leads to the spallation of the ceramic layer resulting in catastrophic failure [40, 41]. Therefore, the TGO layer formation and its progressive thickening should be controlled to minimize the bond coat oxidation induced TBC degradation.

TBCs tend to spall during thermal cycling due to thermal stresses generated for the thermal expansion mismatch with the underneath metal substrate and the temperature gradients, residual stresses during deposition, phase transformations, corrosive and erosive attack of the environment and progressive sintering of the ceramic top coat [42]. The bond coat undergoes phase transformation associated with volume change and develops TGO layer at the interface with the top coat leading to high residual stresses with the increase of operating temperature and long time usage, thereby promoting failure of the TBC system. To overcome these problems, glass-ceramic material was used as bond coat between the yttria stabilized zirconia (YSZ) top coat and the nimonic alloy substrate of the TBC system [43, 44].

TBC system consisting of YSZ ($8Y_2O_3$-$92ZrO_2$) top coat, glass-ceramic bond coat and nickel base superalloy (Nimonic-AE 435) substrate was heated at $1200°C$ for 500 h in air. The approximate glass composition was as follows: SiO_2-40-45; BaO-40-45; CaO-2-6; MgO-2-3; ZnO-2-8; MoO_3-2-8 in wt. % [43]. YSZ ($8Y_2O_3$-$92ZrO_2$) and NiCoCrAlY (Ni-24Co-15Cr-8Al-1Y, in wt.%) applied as the top coat and the bond coat, respectively on the nickel base superalloy (Nimonic-AE 435) substrate was also heat treated under identical heat treatment conditions. The bond coat and top coat thickness was 100±10 μm and 400±15 μm, respectively in the both cases. Both the TBC systems were characterized by SEM and EDX analysis. TGO layer was not observed between the bond coat and the top coat in the glass-ceramic bonded TBC system while the conventional TBC system showed a prominent TGO layer (~ 16 μm thickness) at the interface of the bond coat and the top coat. Therefore, glass-ceramics can be used as oxidation resistant bond coat (Fig. **1**) in thermal barrier coating system [43].

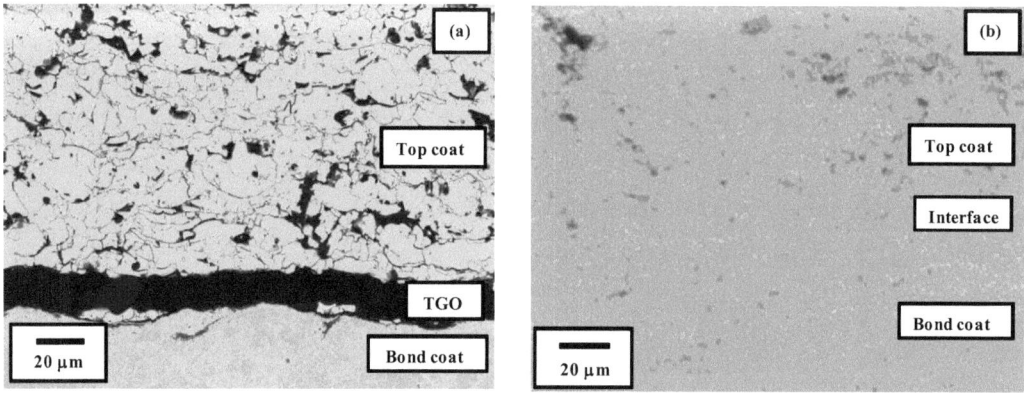

Figure 1: SEM micrographs showing (a) TGO layer between NiCoCrAlY bond coat and YSZ top coat and (b) absence of TGO layer between glass-ceramic bond coat and YSZ top coat [43].

TBC system composed of YSZ ($8Y_2O_3$-$92ZrO_2$) top coat, glass-ceramic bond coat and nimonic alloy (AE 435) substrate was subjected to thermal shock testing from 1000°C to room temperature for 100 cycles. The approximate glass composition was as follows: SiO_2-45; BaO-45; CaO-3; MgO-3; ZnO-2; MoO_3-2 in wt.%. Thickness of the bond coat and top coat was ~100±10 μm and ~ 400±20 μm, respectively [44]. In one type of test, specimens were held at 1000°C for 5 min and then forced air quenched while in the other test specimens were water quenched from the same conditions. XRD and EDX analysis was conducted for phase identification and SEM was used for the microstructural observations. Deterioration was not observed in the top coats after 100 cycles in the case of forced air quenched specimens. On the contrary, the top coats were damaged in the water quenched ones after 100 cycles. After thermal cycling experiments the top coat maintained its phase stability and interfacial crack was not observed at the top coat-bond coat and bond coat-substrate interfaces in both forced air quenched and water quenched specimens [44]. Thus, it can be said that glass-ceramics can be used as thermal shock resistant bond coat in the TBC system.

3.4 Oxidation and Hydrogen Permeation Resistant Glass-Ceramic Coatings for Gamma-Titanium Aluminides by Conventional Processing

Titanium aluminides (TiAl) based on the interdendritic γ-phase (γ-TiAl) are advanced structural materials for high temperature applications in automotive, aerospace and power generation industries [45]. γ-TiAl intermetallic alloys have lower density (3.7-3.9 g/cm^3), high purity (>99.5% purity), higher stiffness (175 GPa at 20°C to 150 GPa at 700°C), higher strength (~650 MPa), improved oxidation and creep resistance up to moderately high temperature, lower coefficient of thermal expansion (CTE) (8.5×10^{-6} at 20°C and 13.75×10^{-6} at 700°C), higher thermal conductivity and higher melting temperature (~1460°C) compared to the conventionally used super alloys [45,46]. However, wider application of γ-TiAl is still limited due to some limiting properties. The major drawbacks are inadequate oxidation resistance at temperatures above 750°C and poor hydrogen permeation resistance [46-49].

MgO-SiO_2-TiO_2 based glass-ceramic coating was applied for the improvement of oxidation resistance of γ-TiAl alloy by vitreous enameling technique [50]. The glass composition selected has been shown in Table **2**.

Results showed that the impervious glass-ceramic coating provided excellent oxidation resistance to γ-TiAl at 800°C even up to 100 h with negligible weight gain (~0.10 mg/cm^2) compared to that (~1.3 mg/cm^2) of the bare alloy (Fig. **2**). The coatings from BaO-MgO-SiO_2, ZnO-Al_2O_3-SiO_2 and BaO-SiO_2 systems also improved the oxidation resistance of the alloy at 800°C up to 100 h. At further higher temperature, *e.g.* 1000°C, MgO-SiO_2-TiO_2 and BaO-SiO_2 based glass-ceramic coatings offered significant protection to the alloy up to 25 h with minimum weight gain (~0.34 mg/cm^2). However, the coating started to peel off from the alloy surface after that period [50].

Table 2: Range of compositions (in weight %) of the coating materials studied [50]

Components	Ni 2/1	Ni 2/2	ABK-13	ABK-103	ARDB
SiO_2	30-40	35-45	30-40	30-40	40-45
BaO	40-50	40-50	-	50-60	-
B_2O_3	3-6	6-9	5-15	-	1-3
CaO	4-5	4-5	-	2-4	-

Table 2: cont....

Al$_2$O$_3$	-	-	15-25	1-2	10-20
MgO	1-2	1-2	10-20	-	1-3
TiO$_2$	-	-	10-20	4-6	-
ZnO	2-3	4-6	-	-	30-40
K$_2$O	7-9	-	5-10	-	-

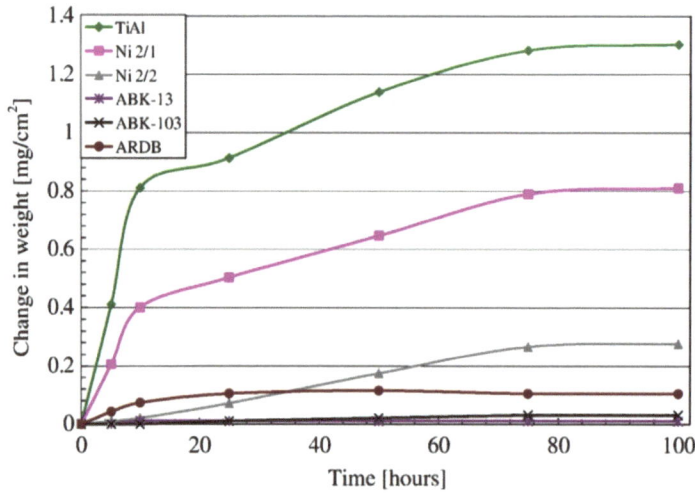

Figure 2: Oxidation weight gain plot for all the coated samples including the bare alloy at 800°C [50].

In the case of high performance applications of many intermetallics hydrogen embrittlement is a major limiting factor. The inward diffusion of hydrogen atoms and formation of brittle hydride phases during hydrogen exposure of the intermetallic components leads to decrease of the component's ductility and fracture toughness. Oxide based glass-ceramic coatings were applied on γ-TiAl by vitreous enameling technique to protect it from hydrogen embrittlement at high temperatures *e.g.* 800 °C at 0.1 MPa gas pressure up to 75 h [51]. It was noted the coated samples were remained mostly unaffected after the test with minimum changes in their microstructures while the uncoated γ-TiAl alloy was severely affected by the hydrogen exposure test. The weight gain of the uncoated alloy after 75 h of hydrogen permeation test was ~1.05 mg/cm^2. On the other hand, the weight gain of the coated samples was only ~0.12 mg/cm^2 and ~0.15 mg/cm^2 for BaO-SiO$_2$-MgO and MgO-SiO$_2$-TiO$_2$ based glass-ceramic coated substrates (Fig. **3**). The coating contained major crystalline phases with little or no trace of the detrimental hydride phases after high temperature exposure of the coated samples in flowing H_2 up to 75 h whereas presence of aluminum hydride and titanium hydride was observed in the uncoated alloy [51].

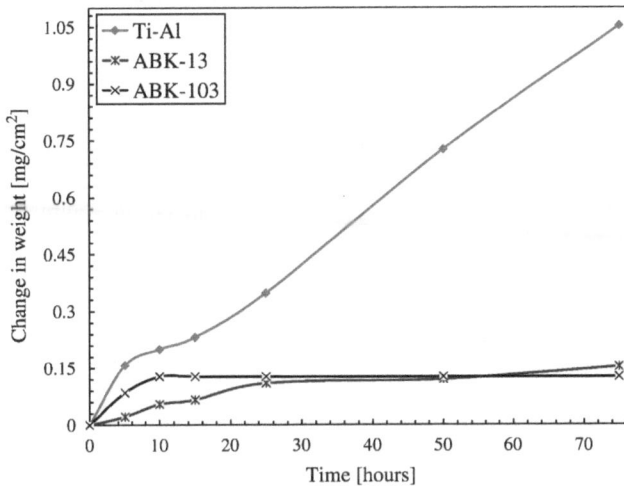

Figure 3: Hydrogen permeation weight gain plot for the coated samples including the uncoated alloy at 800°C [51].

CONCLUSION

The present review demonstrates the synthesis of specialty ceramics and glass-ceramics through exploitation of the microwave and/or conventional processing method. Significant advancement has taken place in the microwave processing of materials in the past decades. However, further research is needed to have a full understanding of this process. Thus, more time may be required to implement the microwave technology for the ceramic industry.

CONFLICT OF INTEREST

There is no conflict of interest with other people or organizations in respect of the present research work.

ACKNOWLEDGEMENTS

The authors are very grateful to Prof. I. Manna, Director, Central Glass and Ceramic Research Institute (CGCRI), Kolkata -700 032, India, for his kind permission to publish this eBook chapter.

REFERENCES

[1]   Samant AN, Dahotre NB. Laser machining of structural ceramics-A review. J Euro Ceram Soc 200; 29: 969-993.

[2]   Pinckneyw LR, Beall GH. Microstructural Evolution in Some Silicate Glass-Ceramics: A Review. J Am Ceram Soc 2008; 91: 773-779.

[3]   Sutton WH. Microwave processing of ceramic materials. Ceram Bull 1989; 68: 376-386.

[4]   Thostenson ET, Chou T-W. Microwave processing: Fundamentals and applications. Composites: Part A 1999; 30: 1055-1071.

[5]   Das S, Mukhopadhyay AK, Datta S, Basu D. Prospects of microwave processing: An overview. Bull Mater Sci 2009; 32: 1-13.

[6]   Das S, Mukhopadhyay AK, Datta S, Basu D. Novel method of developing oxide coating on aluminium using microwave heating. J Mater Sci Lett 2003; 22: 1635-1637.

[7]   Das S, Mukhopadhyay AK, Datta S, Basu D. Aluminium oxide coating by microwave processing. Trans Ind Ceram Soc 2006; 65: 105-110.

[8]   Roy R, Agrawal D, Cheng J, Gedevanishvili S. Full sintering of powder-metal bodies in a microwave field. Nature 1999; 399: 668-670.

[9]   Gedevanishvili S, Agrawal D, Roy R. Microwave combustion synthesis and sintering of intermetallics and alloys. J Mater Sci Lett 1999; 18: 665-668.

[10]   Saitou K. Microwave sintering of iron, cobalt, nickel, copper and stainless steel powders. Scr Mater 2006; 54: 875-879.

[11]   Gerdes T *et al.* In: M F Iskander *et al.* (eds.), Microwave processing of materials V, Materials research society symposium proceedings, Pittsburgh, Pennsylvania, Materials Research Society, 1996, 45.

[12]   Wong WLE, Gupta M. Development of Mg/Cu nanocomposites using microwave assisted rapid sintering. Compos Sci Technol 2007; 67: 1541-1552.

[13]   Zebarjad SM, Sajjadi SA. Microstructure evaluation of Al-Al$_2$O$_3$ composites produced by mechanical alloying method. Mater Des 2006; 27: 684-688.

[14]   Salas O, Ni H, Jayaram V, Vlach KC, Levi CG, Mehrabian R. Nucleation and growth of Al$_2$O$_3$/metal composites by oxidation of aluminum alloys. J Mater Res 1991; 6: 1964-1981.

[15]   Vlach KC, Salas O, Ni H, Jayaram V, Levi CG, Mehrabian R. A thermogravimetric study of the oxidative growth of Al$_2$O$_3$/Al alloy composites. J Mater Res 1991; 6: 1982-1995.

[16]   Ng DHL, Zhao Q, Qin C, Ho MW, Hong Y. Formation of aluminum/alumina ceramic matrix composite by oxidizing an Al-Si-Mg alloy. J Euro Ceram Soc 2001; 21: 1049-1053.

[17]   Guillard FJAH, Hand RJ, Lee WE. Br Ceram Trans 1994; 93: 129-136.

[18]   Wu CML, Han GW. Synthesis of an Al$_2$O$_3$/Al co-continuous composite by reactive melt infiltration. Mater Charact 2007; 58: 416-422.

[19]   Chou SN, Huang JL, Lii DF, Lu HH. The mechanical properties and microstructure of Al$_2$O$_3$/aluminum alloy composites fabricated by squeeze casting. J Alloy Compd 2007; 436: 124-130.

[20]   Chou SN, Lu HH, Lii DF, Huang JL. Processing and physical properties of Al$_2$O$_3$/aluminum alloy composites. Ceram Int 2009; 35: 7-12.

[21]   Hoseini M, Meratian M. Fabrication of *in situ* aluminum-alumina composite with glass powder. J Alloy Compd 2009; 471: 378-382.

[22]   Shafiei-Zarghani A, Kashani-Bozorg SF, Zarei-Hanzaki A. Microstructures and mechanical properties of Al/ Al$_2$O$_3$ surface nano-composite layer produced by friction stir processing. Mat Sci Eng A 2009; 500: 84-91.

[23] Rahimian M, Parvin N and Ehsani N. Investigation of particle size and amount of alumina on microstructure and mechanical properties of Al matrix composite made by powder metallurgy. Mat Sci Eng A 2010; 527: 1031-1038.

[24] Tatar C, Özdemir N. Investigation of thermal conductivity and microstructure of the α-Al_2O_3 particulate reinforced aluminum composites (Al/Al_2O_3-MMC) by powder metallurgy method. Physica B 2010; 405: 896-899.

[25] Das S, Basu D, Datta S, Mukhopadhyay AK. A process of making Al-Al_2O_3 composites useful in engineering applications. Indian patent no. 233814.

[26] Das S, Datta S, Mukhopadhyay AK, Pal KS, Basu D. Al-Al_2O_3 core-shell composite by microwave induced oxidation of aluminium powder. Mater Chem Phys 2010; 122: 574-581.

[27] Ghosh S, Pal KS, Dandapat N, Mukhopadhyay AK, Datta S, Basu D. Characterization of microwave processed aluminium powder. Ceram Int 2011; 37: 1115-1119.

[28] Das S, Mukhopadhyay AK, Datta S, Dandapat N, Basu D. Effect of dopants on the phase formation in microwave processed Al-Al_2O_3 composites. J Alloy Compd 2010; 500: 231-236.

[29] Das S, Mukhopadhyay AK, Datta S, Basu D. Microwave sintering of titania. Trans Ind Ceram Soc 2005; 64: 143−148.

[30] Datta S, Das S. A new high temperature resistant glass-ceramic coating for gas turbine engine components. Bull Mater Sci 2005; 28: 689−696.

[31] Datta S, Das S. A new high temperature resistant glass-ceramic coating developed in CGCRI, Kolkata. Trans Ind Ceram Soc 2005; 64: 25−32.

[32] Cozzi AD, Fathi Z, Clark DE. Crystallization of sol-gel derived barium aluminosilicate in a 2.45 GHz microwave field. In: Clark DE *et al.* (eds.), Ceram Trans. Westerville, Ohio, The Am Ceram Soc Inc 1993; pp 317−324.

[33] Siligardi C, Leonelli C, Fang Y and Agrawal D. Modifications on bulk crystallization of glasses belonging to M_2O-CaO-SiO_2-ZrO_2 system in a 2.45 GHz microwave field. In: M F Iskander *et al.* (eds.), Microwave processing of materials V, Pennsylvania, Pittsburgh, Materials Research Society, Mater Res Soc Symp Proc 1996; pp 429-434.

[34] Siligardi C, Leonelli C, Bondioli F, Corradi A, Pellacani GC. Densification of glass powders belonging to the CaO-ZrO_2-SiO_2 system by microwave heating. J Euro Ceram Soc 2000; 20: 177−183.

[35] Das S, Mukhopadhyay AK, Datta S, Das GC, Basu D. Hard glass-ceramic coating by microwave processing. J Euro Ceram Soc 2008; 28: 729-738.

[36] Das S, Basu D, Datta S, Mukhopadhyay AK. Crystallization of glass coating by microwave heating. Trans Ind Ceram Soc 2008; 67: 139−146.

[37] Das S, Mukhopadhyay AK, Datta S, Basu D. Evaluation of microwave processed glass-ceramic coating on nimonic superalloy substrate. Ceram Int 2010; 36: 1125−1130.

[38] Mao WG, Dai CY, Zhou YC, Liu QX. An experimental investigation on thermo-mechanical buckling delamination failure characteristic of air plasma sprayed thermal barrier coatings. Surf Coat Technol 2007; 201: 6217-6227.

[39] Taymaz I. The effect of thermal barrier coatings on diesel engine performance. Surf Coat Technol 2007; 201: 5249-5252.

[40] Martena M, Botto D, Fino P, Sabbadini S, Gola MM, Badini C. Modelling of TBC system failure: Stress distribution as a function of TGO thickness and thermal expansion mismatch. Eng Fail Anal 2006; 13: 409-426.

[41] Chen WR, Wu X, Marple BR, Patnaik PC. Oxidation and crack nucleation/growth in an air-plasma-sprayed thermal barrier coating with NiCrAlY bond coat. Surf Coat Technol 2005; 197:109-115.

[42] Nusair Khan A, Lu J. Behavior of air plasma sprayed thermal barrier coatings, subject to intense thermal cycling. Surf Coat Technol 2003; 166: 37-43.

[43] Das S, Datta S, Basu D, Das GC. Glass-ceramics as oxidation resistant bond coat in thermal barrier coating system. Ceram Int 2009; 35: 1403−1406.

[44] Das S, Datta S, Basu D, Das GC. Thermal cyclic behavior of glass-ceramic bonded thermal barrier coating on nimonic alloy substrate. Ceram Int 2009; 35: 2123−2129.

[45] Clemens H, Kestler H. Processing and Applications of Intermetallic γ-TiAl-Based Alloys. Adv Eng Mater 2000; 2: 551−570.

[46] Voice WE, Hendersonb M, Shelton EFJ, Wu X. Gamma titanium aluminide, TNB. Intermetallics 2005; 13: 959−964.

[47] Estupiñan HA, Uribe I, Sundaram PA. Hydrogen permeation in gamma titanium aluminides. Corr Sci 2006; 48: 4216−4222.

[48] Loria EA. Gamma titanium aluminides as prospective structural materials. Intermetallics 2000; 8: 1339−1345.

[49] Vaidya RU, Park YS, Zhe J, Gray III GT, Butt DP. High-Temperature Oxidation of Ti-48Al-2Nb-2Cr and Ti-25Al-10Nb-3V-1Mo. Oxid Met 1998; 50: 215−240.

[50] Sarkar S, Datta S, Das S, Basu D. Oxidation protection of gamma-titanium aluminide using glass-ceramic coatings. Surf Coat Technol 2009; 203: 1797−1805.

[51] Sarkar S, Datta S, Das S, Basu D. Hydrogen permeation resistant glass-ceramic coatings for gamma-titanium aluminide. Surf Coat Technol 2009; 204: 391−397.

Send Orders of Reprints at bspsaif@emirates.net.ae

CHAPTER 4

Development and Characterization of Lanthanides Doped Hydroxyapatite Composites for Bone Tissue Application

João Coelho[1], Sooraj H. Nandyala[1,2,*], Pedro S. Gomes[3], Mónica P. Garcia[3], Maria A. Lopes[4], Maria H. Fernandes[3] and José D. Santos[4]

[1]*INESC Porto, Rua do Campo Alegre, 687, 4169-007 Porto, Portugal;* [2]*Departamento de Física, Faculdade de Ciências, Universidade do Porto, Rua do Campo Alegre, 687, 4169-007 Porto, Portugal;* [3]*Laboratório de Farmacologia e Biocompatibilidade Celular. Faculdade de Medicina Dentária, Universidade do Porto, Rua Dr. Manuel Pereira da Silva, 4200-393 Porto, Portugal and* [4]*DEMM, Faculty of Engineering, University of Porto, Rua Dr. Roberto Frias, 4200-465 Porto, Portugal*

Abstract: This work reports the preparation and characterization of newly developed $10CaF_2$-$10Na_2CO_3$-$15CaO$-$59P_2O_5$-$5SiO_2$ glasses, doped with lanthanides, in this case cerium and lanthanum oxide ($10CaF_2$-$10Na_2CO_3$-$15CaO$-$59P_2O_5$-$1CeO_2$-$5SiO_2$ and $10CaF_2$-$10Na_2CO_3$-$15CaO$-$59P_2O_5$-$1La_2O_3$-$5SiO_2$, respectively). The structure and morphology of the developed glasses have been investigated by Raman and FTIR spectroscopy. Scanning electron microscopy with an energy dispersive analyzer and X-ray mapping was used to assess the morphological properties of the glasses. Glass-ceramic composites, for bone tissue applications, were obtained by the mixture of 2.5wt% of each glass with 97.5wt% of hydroxyapatite. These were also analyzed by means of XRD and SEM. Composites were biologically evaluated with human osteoblastic-like cells. Lanthanide doped-hydroxyapatite composites revealed an improved biological behaviour, regarding cell adhesion and proliferation, compared to hydroxyapatite and undoped glass-hydroxyapatite composites. Lanthanide doped composites reported an adequate biocompatibility, further enhancing the cell adhesion and proliferation, behaviour that indicates a prospective application in bone tissue engineering.

Keywords: Lanthanide, glass-hydroxyapatite composites, cell culture, biocompatibility, bone tissue, biological characterization, biocompatibility, bone tissue engineering, cytocompatibility, cell culture, cerium oxide doped HA composites, doped glasses, glass-hydroxyapatite composites, hydroxyapatite, *in vitro* testing, Lanthanum oxide doped HA composites, materials characterization, osteoblasts.

1. INTRODUCTION

Nowadays, there is a great effort in the development and production of novel

---

*Address correspondence to Sooraj H. Nandyala: FCUP/INESC Porto/Department of Physics, Faculty of Sciences, University of Porto, Rua do Campo Alegre, 687, 4169-007 Porto, Portugal; Tel: +351 22 0402302 and Fax: +351 22 0402 437; Email: nandyala.sooraj@fc.up.pt

Sooraj H. Nandyala and José D. Santos (Eds)
All rights reserved-© 2013 Bentham Science Publishers

scaffolds for bone applications. These scaffolds are also expected to substitute the shape and volume of any lost tissue, provide biomechanical support during the healing process and promote the formation of a direct bond between the implant and the remaining tissue [1]. Hydroxyapatite-based biomaterials have been developed and used for a long time, with in the clinical scenario. As an abundant element in the body, hydroxyapatite (HA) does not present any kind of cytotoxicity or antigenic behaviour [2, 3]. However, stoichiometric HA ceramics, do not present a microstructure similar to that of the natural bone and their biomechanical properties are broadly poor [4-7]. For instance, the fracture toughness (K_{Ic}) of pure HA is around $1 MPam^{1/2}$, while the reference value for human bone is up to $12 MPam^{1/2}$ [8-10]. Therefore, the use of stoichiometric HA ceramics is limited to implant coatings and non-load-bearing applications [4, 7, 11]. The slow dissolution of these materials is other challenging problem within the clinical field, impairing the timely ingrowth of neighbouring bone tissue [6]. The dissolution of HA is governed by various factors, such as the grain size, morphology, surface area, chemical composition, crystal structure, crystallinity, and micro-porosity [4, 12]. In order to improve the reported biomechanical and dissolution issues, the addition of a second phase to the HA matrix has been assayed [8, 9, 13]. Strategies that have been developed include the addition of glasses to the HA matrix, which results in the formation of glass-reinforced HA (GR-HA) [8, 10]. These composites exhibit several ionic substitutions within the HA matrix [2]. Under physiological conditions, many of these trace elements are already present in the natural matrix of bone, and are known to play an important role in bone metabolic regulation [14, 15]. Within developed biomaterials for bone tissue regeneration, the presence of Zn^{2+}, Mg^{2+}, Sr^{2+}, Ba^{2+}, $(CO_3)^{2-}$, $(SiO_4)^{4-}$ and Ga^{3+}, added to the HA matrix, has been assayed and proven to enhance the biological behaviour of the composite, as assessed by *in vitro* and *in vivo* testing [2, 5, 14-19].

More recently, HA substituted with rare-earth (RE) ions became the focus of interest for bone related biomaterial applications [20]. RE elements (or lanthanides) are the family of elements from the lanthanum (Z=57) to lutetium (Z=71). Because of their resemblance to calcium, lanthanides exhibit a pronounced biological activity, as they are able to replace Ca^{2+} in structured molecules [21-23]. Regarding the biomedical field, lanthanides have been mainly used in two distinct applications: as contrast agents for magnetic resonance or as luminescent probes for biosensors within *in vivo* imaging applications [21, 22, 24]. They exhibit a large Stokes shift, narrow emission bands and long emission lifetime, ranging from microseconds to milliseconds, which make them suitable particularly for biomedical optical or imaging applications [25, 26].

In the past few years, lanthanides have been found to play a role in bone metabolic equilibrium. Li *et al.* reported that a long-term oral supplementation of lanthanum in the rats (at a low dose) caused lanthanum accumulation in the bone tissue, reduced the bone Ca/P ratio, decreased bone density, induced changes in the microstructure of bone and increased bone crystalinity [27]. Barta *et al.* have investigated the use of lanthanides for the management of bone resorption disorders, showing that lanthanides can replace calcium in bones and cause local activation of osteoblasts, which are responsible for bone formation [28]. Additionally, lanthanides have been shown to be effective within the management of bone-related inflammatory conditions. They seem to be able to modulate the inflammatory process in rheumatoid arthritis and osteoarthritis, by inhibiting the activity of matrix metalloproteinases, such as collagenase [21].

Several research applications relating to *in vitro* methodologies have addressed the effect of lanthanides in cellular populations with relevance to bone metabolism. A short term biocompatibility assessment of Sm_2O_3 has indicated its biosafety, with no damaging cytotoxic effects or loss of biofunctionality by a human osteosarcoma cell line (HOS cells) [29]. Quarles *et al.* found that Gd^{3+} could stimulate DNA synthesis in MC3T3-E1 osteoblasts *in vitro*, in a dose-dependent fashion, in a process probably mediated by the activation of membrane associated G-proteins [30]. Zhan *et al.* reported that La^{3+}, at 1×10^{-5} mol.L^{-1} induced the osteogenic differentiation of primary osteoblasts, as assessed by the significantly increased activity of alkaline phosphatase up to 3 folds [31]. Wang *et al.* also revealed that La^{3+} was able to enhance the osteoblastic differentiation process, in rat calvaria-derived osteoblasts [32]. Ce^{3+}, at concentrations between 1×10^{-9} and 1×10^{-4} mol.L^{-1} promoted the proliferation of mouse osteoblasts. Interestingly, low concentrations (between 1×10^{-9} and 1×10^{-7} mol.L^{-1}) inhibited the formation of a mineralized extracellular matrix, while high concentrations (between 1×10^{-6} and 1×10^{-4} mol.L^{-1}) induced the cell-mediated mineralization process [33]. The effect of lanthanides on osteoclastic function has also been addressed. La^{3+} was shown to promote or inhibit the formation and bone-resorbing activity of rabbit osteoclast-like cells depending on the assayed concentration [34]. Additionally, effects of other rare earth ions (Nd^{3+}, Gd^{3+}, Dy^{3+}, Sm^{3+} and Er^{3+}) on rabbit osteoclastic function were also found to be bidirectional, depending on their concentration [35].

Because of their ability to modulate the bone regeneration process, the inclusion of RE elements on the composition of biomaterials for bone tissue regeneration has been assessed. In some applications lanthanides have been included within the HA matrix, in order to enhance its bioactivity. Promising results have been attained, namely regarding the enhancement of the osteoblastic behaviour, *in vitro* [36-39]. Alternative approaches include the development of bioactive glasses containing lanthanides which, when added to the HA matrix, allow for the development of a bioactive composite material with improved biomechanical properties and the ability to release, in a controlled way, bioactive species, including lanthanides. Hence, this work reports the preparation and characterization of $10CaF_2$-$10Na_2CO_3$-$15CaO$-$59P_2O_5$-$5SiO_2$-$1Ln$ glasses (Si-P_2O_5 glasses), in which Ln stands for lanthanide, in particular for this study cerium and lanthanum oxide. The structure and morphology of the glasses have been investigated by Fourier transform infrared spectroscopy (FTIR), Raman, and scanning electron microscopy (SEM) with an energy dispersive analyzer (EDS). Glass-ceramic composites (GR-HA) were obtained by the mixture of 2.5wt% of each glass with 97.5wt% of hydroxyapatite (HA). These were also analyzed by means of XRD and SEM. Additionally, a preliminary biological characterization of the developed composites was conducted with human osteoblastic-like cells.

2. MATERIALS AND METHODS

2.1. Fabrication of the Glasses

Phosphosilicate glasses have been prepared by the quenching technique. Briefly, they were produced by melting the mixture of analytical grade CaF_2, Na_2CO_3, CaO, P_2O_5, SiO_2, CeO_2 and La_2O_3 (Sigma Aldrich, 99.99% purity) in crucibles, for about an hour in an electrical furnace at temperature of 1000 °C. Undoped (Si) and lanthanide-doped glasses (Ce and La) were produced with the following chemical composition:

- Si: $10CaF_2\text{-}10Na_2CO_3\text{-}15CaO\text{-}60P_2O_5\text{-}5SiO_2$

- Ce: $10CaF_2\text{-}10Na_2CO_3\text{-}15CaO\text{-}59P_2O_5\text{-}1CeO_2\text{ -}5SiO_2$

- La: $10CaF_2\text{-}10Na_2CO_3\text{-}15CaO\text{-}59P_2O_5\text{-}1La_2O_3\text{-}5SiO_2$

The glasses obtained were circular in design, 2–3 cm in diameter, a thickness of 0.33-0.38 cm and presented good transparency. After, the glasses were annealed at 200°C for an hour to remove thermal strains. For in-depth characterization and preparation of GR-HA composites, the glasses were crushed and sieved to produce granules (≤ 75 μm).

2.2. Preparation of the Hydroxyapatite (HA) and GR-HA Composites

Hydroxyapatite (HA) is a naturally occurring mineral with the chemical formula $Ca_{10}(PO_4)_6(OH)_2$. HA can be obtained from natural sources (*i.e.* from corals or mammal's bones, for instance), or can be chemically synthesised. In this work, HA was prepared by the addition of phosphoric acid to a saturated solution of calcium hydroxide, as demonstrated in the following equation (Eq. 1):

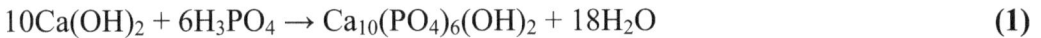

$$10Ca(OH)_2 + 6H_3PO_4 \rightarrow Ca_{10}(PO_4)_6(OH)_2 + 18H_2O \tag{1}$$

This technique allows the preparation of high purity HA, if the starting materials also present a high purity and the preparation conditions are optimized. In brief, analytical grade calcium hydroxide, $Ca(OH)_2$ and ortho phosphoric acid, H_3PO_4 (Sigma Aldrich, 99.99% purity), have been independently mixed with deionized water and subsequently, $Ca(OH)_2$ has been slowly transferred into the H_3PO_4, over a period of 4 hours, under constant mixing conditions (100 rpm), with the pH maintained at 10.50. After 24 hours, the material was filtered to remove the excess of water and dried at 60°C for approximately 72 hours. The dried product was finally crushed and sieved to produce granules (≤ 75 μm). HA disks were prepared by isostatic pressure at 200 MPa and sintered at 1300°C for 1 hour. These disks were used as a control substrate in the biological characterization of GR-HA with human osteoblastic-like cells. Prior to cell seeding, sample discs were sterilized in a steam autoclave.

GR-HA composites were obtained by mixing 2.5wt% of each glass granules (Si, Ce and La glasses) with 97.5wt% of HA granules, in deionized water, with turbular mixing to assure homogeneity. The composites obtained (Si-HA, Ce-HA and La-HA, respectively) were dried at 60°C for 24 hours and sieved to select granules, of less than 75 μm in size. Composite disks were prepared as previously described for HA.

2.3. Characterization Techniques of Glasses and GR-HA Composites

Glass density was measured using ethyl glycol as an immersion liquid by Archimedes' principle on Mettler Toledo balance. X-ray diffraction (XRD) was performed on powder samples of both glasses and composites by using Siemens D 5000 diffractometer with Cu-Kα radiation ($\lambda= 1.5418$Å). The scans were made in the range of 25-40° (2θ) with a step size of 0.02° and a count time of 2 sec/step.

Fourier transform infrared (FTIR) spectra were recorded in KBr pellets in the range of 400-4000cm^{-1} at a resolution of 4 cm^{-1} on a Jasco FT/IR – 460 PLUS spectrophotometer. The spectrum was recorded by using the KBr pellets containing approximately 95% of KBr and 5% of glass sample in 250mg of a pellet. Before recording the spectra, the samples were dried in a Buchi Glass Oven- B-585, at 120°C, for 1 hour.

SEM imaging and colour coded X-ray mapping were performed in a FEI Quanta 400 FEG ESEM/EDAX Genesis X4M, a high resolution environmental Scanning Electron Microscope with X-ray microanalysis and backscattered electron diffraction pattern analysis. Both SEM images and maps were obtained in high vacuum mode and with an acceleration of 500kV. In order to avoid superficial charge accumulation, the samples were covered with a carbon film. X-ray maps were acquired in regions of 500x400μm for a period of 15 minutes.

The unpolarized micro-sampling Raman spectra of the glasses have been recorded in the backscattering geometry, at room temperature, by using an Olympus BH2 UMA microscope and a x50 lens. The 514.53 nm polarized line of a Ar^{+} laser was used for excitation, with an incident power of about 150 mW impinging the sample. The scattered light was analyzed using a T64000 Jobin-Yvon spectrometer operating in the triple subtractive mode, and equipped with a LN$_2$ cooled CCD. Identical conditions were kept constant for all measurements. The spectral slit width was about 1.5 cm^{-1} and the spatial resolution on the sample was about 1μm.

2.4. Biological Characterization of GR-HA Composites with MG63 Osteoblast-like Cells

MG63 cells were cultured in α-Minimal Essential Medium (α-MEM) containing 10% fetal bovine serum, 50 $\mu g.ml^{-1}$ ascorbic acid, 50 $\mu g.ml^{-1}$ gentamicin and 2.5 $\mu g.ml^{-1}$ fungizone, at 37°C, in a humidified atmosphere of 5% CO_2 in air. For sub culturing, the cell monolayer (at around 70-80% confluence) was washed twice with phosphate-buffered saline (PBS) and incubated with trypsin – EDTA solution (0.05% trypsin, 0.25% EDTA), for 5 minutes, at 37°C, to promote cell detachment. The effect of trypsin was following inhibited by the addition of complete culture medium, at 37°C. Cells were then re-suspended in complete culture medium and cultured (2×10^4 cells.cm^{-2}) for 5 days on the surface of HA and GR-HA composites. Sample materials were previously incubated in complete culture medium for 30 minutes. The medium was changed every 2-3 days. Established cultures were evaluated at days 2 and 5 for cell viability/proliferation (MTT assay) and observed by confocal laser scanning microscopy (CLSM).

MTT assay was used to estimate cell viability/proliferation. This assay is based on the reduction of 3-(4, 5-dimethylthiazol-2-yl)-2,5-diphenyltetrazolium bromide to a purple formazan product by viable cells. At each endpoint, cultured cells were incubated with 0.5 mg/ml of MTT (in culture medium) during the last 4 h of the culture (at 37°C, in a humidified atmosphere of 95% air and 5% CO_2). The medium was then decanted and samples were photographed using a stereo microscope (Nikon SMZ445) and a CCD camera (Nikon DS-2M). Following image acquisition, the stained product was dissolved with dimethylsulphoxide and the absorbance was determined in an ELISA reader (Biotek Synergy II) at 600 nm. Results were expressed as absorbance per square centimetre (A.cm^{-2}). Analysis of the results was carried out with one-way analysis of variance (ANOVA), with a significance level of $p \leq 0.05$.

For the CLSM observation, cells grown over the samples were fixed, at the selected end points, in 3.7% methanol-free formaldehyde. Following, cells were permeabilized with 0.1% Triton, and incubated with bovine serum albumin (Sigma Aldrich) at 10 mg.ml^{-1} in PBS for 1 hour, in order to block non-specific interactions. The cell's cytoskeleton filamentous actin (F-actin) was visualized by treating permeabilised cells with Alexa Fluor® 488-conjugated phalloidin dye (Invitrogen), at a concentration of 1:100 in PBS, for 20 minutes. Cells were also counterstained (t=8 min) with propidium iodide (Sigma Aldrich) at 10 mg.ml^{-1} for cell nuclei labelling. Stained seeded samples were mounted in Vectashield® (Vector laboratories) and examined in a Leica SP2 AOBS (Leica Microsystems®) microscope.

3. RESULTS AND DISCUSSION

3.1. Glasses Characterization

3.1.1. Density and Molar Volume of the Prepared Glasses

Table **1** shows the some measured physical properties of the prepared glasses. It is known that the addition of a dopant to a glass may promote changes in the glass network that can affect its physical properties.

Table 1: Physical properties of the prepared glasses

Glasses	Density (d) (g.cm^{-3})	Average Molecular weight ($\overline{M}$) (g/mol)	Molar Volume (V$_m$) (cm^3.mol^{-1})	Ion concentration N (x10^{20} ions/cm^3)	Refractive index (n)	r$_i$ (nm)
Si	2.548	114.985	45.116	--	1.625	0.15039
Ce	2.569	115.287	44.872	3.44	1.609	0.15107
La	2.500	116.824	46.721	3.22	1.623	0.14771

The variation of molar volume of a glass can be explained by the following equation:

$$\rho = \frac{M}{V} \tag{2}$$

where M stands for the average molecular weight of the oxide glass, and ρ for density of the glass. The structure of phosphosilicate glasses is much complex and is well described in the Q^n terminology. The PO_4^{3-} PO_4^{3-} tetrahedron can be linked to n bridging oxygen atoms (BO), which will determine the phosphate species present in the glass (n=0 to 3). Orthophosphate units correspond to n=0 and pyrophosphates to n=1. If n=2, the glass is composed of very long chains or rings of metaphosphates, and for n=3, the glass presents a three-dimensional structure of ultra-phosphates [40]. Therefore, the structure of these glasses is usually referred as Q^n, with the number n relating to BO. Furthermore, the P_2O_5-SiO_2 system is being intensively studied in the biomedical field since these glasses present a remarkable bioactivity [40-42].

The density d and average molecular weight $\overline{M}$ were also used to evaluate the concentration of dopant ions (N X 10^{20} ions /cm^3), with the following equation and data presented in Table **1**.

$$N_{DI}(cm^{-3}) = \frac{\frac{n(DI)}{n_t} \times N_A \times \rho_g}{M} \tag{3}$$

In this equation (Eq. 3) the *n(DI)* is the number of moles of dopant ions in the batch, n_t the total number of moles in the glass, N_A the Avogadro's number, ρ_g the density of the glass and M, its average molecular weight [43].

3.1.2. FTIR and Raman Studies of the Prepared Glasses

The spectra of the prepared glasses are presented in Fig. **1**. It is possible to observe four main regions in the spectra, *i.e.* 400-600 cm^{-1}, 600-800 cm^{-1}, 800-1200 cm^{-1} and 1200-1400 cm^{-1}. The region corresponding to 1200-1400 cm^{-1} originates from the vibrations of the non-bridging oxygen atoms associated to phosphorus atoms and P=O bonds' vibrations [44, 45]. The broad band in the region corresponding to 800-1200 cm^{-1} can be attributed to the superposition of Si-O and P-O stretching vibrations [46, 47]. The bands around the 600-800 cm^{-1} region are assigned to the superposition of the Si-O-Si, Si-O-P and P-O-P vibrations [46, 47]. Finally, the region corresponding to 400-600 cm^{-1} is the usual range of P-O vibrational modes. In the literature it is noted that this region is more pronounced when Ca^{2+} ions are present in the matrix, as a glass modifier component.

In the spectra shown, the 400-600 cm^{-1} region has very strong bands, confirming the modifying behaviour of calcium ions [47]. Moreover, the addition of cerium and lanthanum oxide to the glass system also induces some changes in the structure of the glasses. Near 875 cm^{-1} a shoulder is visible, especially for the sample doped with lanthanum. Furthermore, the intensity of the bands changes drastically in the three assayed glasses, which indicates that cerium and lanthanum oxide behave as glass modifiers. In Fig. **2**, the deconvolution of the spectra of the Si glasses is shown, revealing the main bands of these glasses.

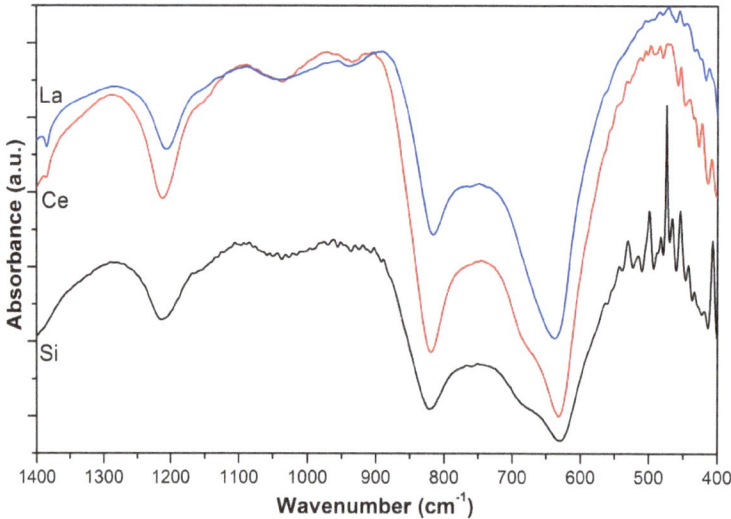

Figure 1: FTIR absorbance spectra of the Si, and Si-doped Ce and La glasses.

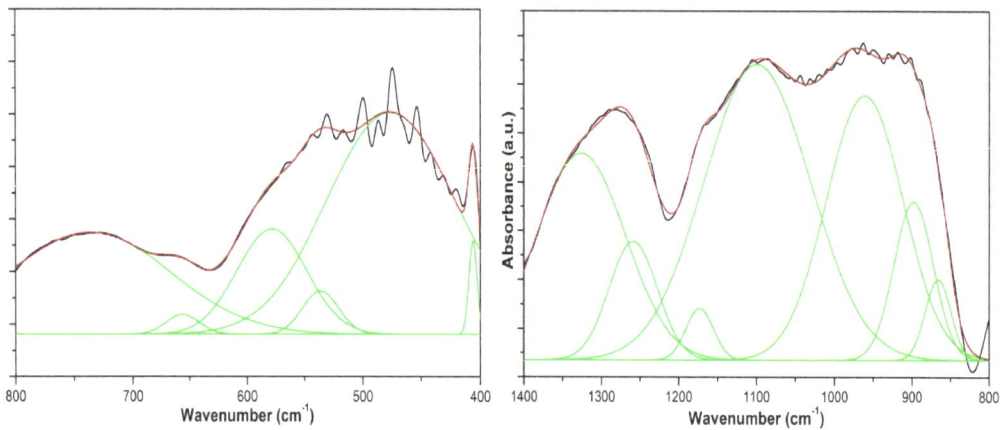

Figure 2: Deconvoluted FTIR spectra of Si glasses in different wavelengths.

Table **2** gives the assignment of the deconvoluted bands in the prepared glasses. Since the band positions are different for the three glasses, this can be indicatative the modifying behaviour of the dopants.

Table 2: FTIR bands assignment and their position in the prepared glasses

Wavenumber (cm^{-1})	Assignments	Glasses		
		Si	Ce	La
405 [48]	O-P-O bending	406	423	405
472 [48]		475	--	454
450 [49]	La-O stretching vibrations	--	--	--
460 [46]	Rocking Si–O–Si vibration	--	--	--
466 [50]	Ce-O stretching vibration	--	--	--
477 [48]	Si-O-Si bending	536	511	--
522 [48]		578	567	567
695 [48]	Six membered rings (Si)	656	667	697
781 [48]	Bending Si–O–Si vibration	786	781	781
800 [46, 48]		--	--	--
915 [48]	Stretching vibration of non-bridging oxygen (Si-O⁻)	898	895	880
944 [48]		--	--	--
950 [51]	Si-O-Ca vibration	960	965	967
995 [48]	P-O symmetric stretching	--	--	--
1062 [48]	Si-O-Si asymetric stretching vibrations	--	--	--
1127 [48]		--	--	--
1160 [48]		--	--	--
1076 [48]	P-O asymmetric stretching	--	--	--
1134 [48]		1100	1102	1114
1170 [48]		1174	1172	1170
1240-1260 [44]	Asymmetric stretching vibrations of non-bridging oxygens bonded to phosphorus	1257	1259	1250
1330 [45]	P=O bonds	1327	1339	1324

In order to confirm these results, Raman analysis was also performed and this is shown in Fig. **3** for the three prepared glasses. It is possible to observe three main regions; 600-800 cm^{-1}, 1100-1250cm^{-1} and 1250-1400 cm^{-1}. These regions can be assigned, respectively, to the vibrations of P-O-P bonds, symmetric vibration of the middle chain (PO$_2$)$^-$ and vibrations of the P-O double bond [52]. No large shift in the peak centre is visible, however their relative intensity decreases with the addition of cerium and lanthanum oxide. In pure P$_2$O$_5$ glasses, there are two main bands at 1390 cm^{-1} and 640 cm^{-1}. The first band is assigned to the P=O vibrations and it can shift by around 130 cm^{-1} with the addition of alkali oxides to the sample

[53]. This outcome is in good agreement with the bands observed in the spectra displayed in Fig. **3**. This is the result of the delocalization of bonds in Q^3 tetrahedra, from the P=O terminal oxygen bond to the bridging P-O-P oxygen bonds [53]. The bond at around 1180 cm^{-1} is also characteristic of alkali oxides added to P_2O_5 glasses and it increases with the oxide content [53].

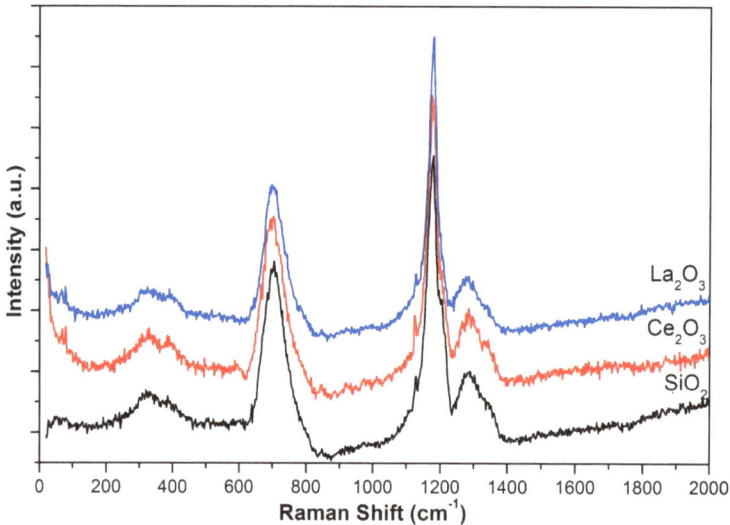

Figure 3: Raman spectra of the prepared glasses.

In Table **3**, the centre position of bands for the main assigned peaks is presented. These results are in good agreement with FTIR data and literature reports, as identified in the table. Moreover, as verified with infrared analysis, cerium and lanthanum were not detected.

Table 3: Raman peaks assignment for the $5SiO_2$-P_2O_5 and doped glasses

Wavenumber (cm^{-1})	Assignments	Glasses		
		Si	Ce	La
		5%	5%	5%
340–350 [46, 54]	Bending vibrations of phosphate polyhedra	343	336	341
330 [55]	La-O vibrations	--	--	--
464 [56]	Ce-O vibrations	--	--	--
560-660 [57]	Si rocking vibrations			
680 [52]	Symmetric vibrations of P-O-P bonds	704	701	703
900–970 [57]	non-bridging silicon–oxygen bond			

Table 3: cont....

970–1100 [58]	Symmetric and anti-symmetric P-O stretching vibrations			
1064-1183 [46]	Si-O-Si stretching vibrations	1123	1175	1065
1180 [52]	Symetric vibration of the middle chain $(PO_2)^-$	1178	1176	1176
1270 [52, 54]	Vibrations of the P-O double bond	1283	1286	1276

3.1.3. Morphological and Structural Analysis of Prepared Glasses

In Fig. **4**, a representative high magnification SEM micrograph of the prepared Ce glass is presented. This image shows an organised and homogenously smooth glassy surface, similar to those obtained in the observation of the Si and La glasses. X-Ray mapping (data not shown) of these samples did not reveal grains or organised clusters of dopants, which suggests a homogeneous distribution of ions within the glass matrix.

Figure 4: Representative SEM micrograph of the $5SiO_2$-P_2O_5 glass doped with cerium (Ce glass).

Regarding EDS analysis (Fig. **5**), the assayed samples of the three glasses presented the expected theoretical composition. Accordingly, because of their reduced concentration, the dopant ions are barely detected.

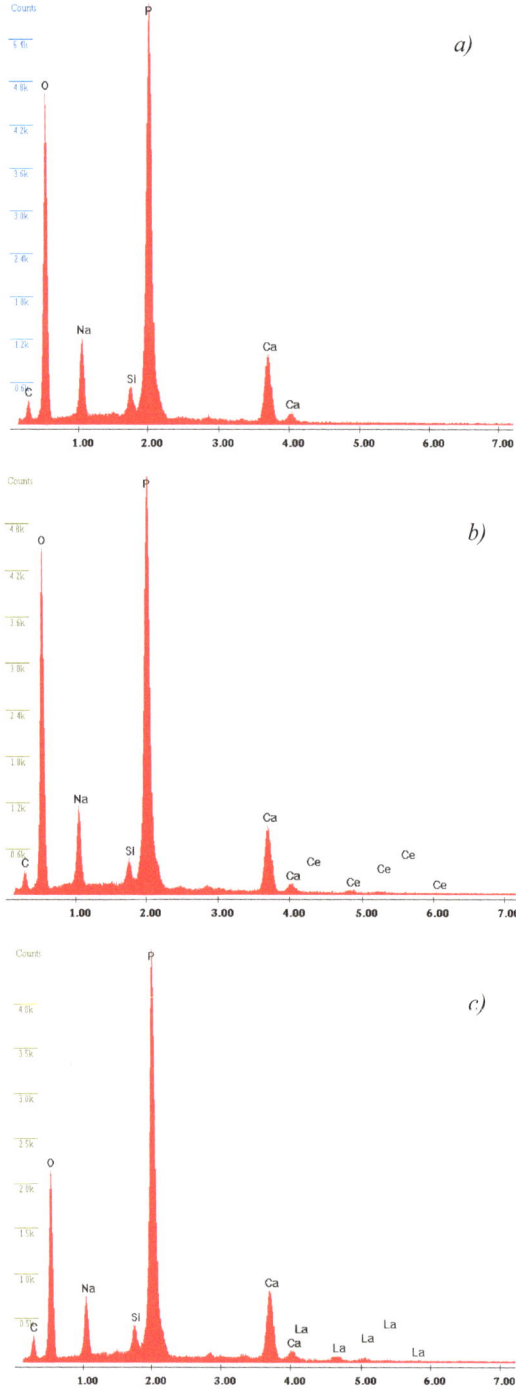

Figure 5: Representative EDS spectra of the prepared glasses. (a) Si glass; (b) Ce glass; and (c) La glass.

3.2. Characterization of GR-HA Composites

3.2.1. Morphological and Structural Analysis of GR-HA Composites

SEM micrographs of the Si-HA, Ce-HA and La-HA composite samples revealed a similar structure. Representative SEM micrographs of the composites and respective EDS spectra are presented in Fig. **6**.

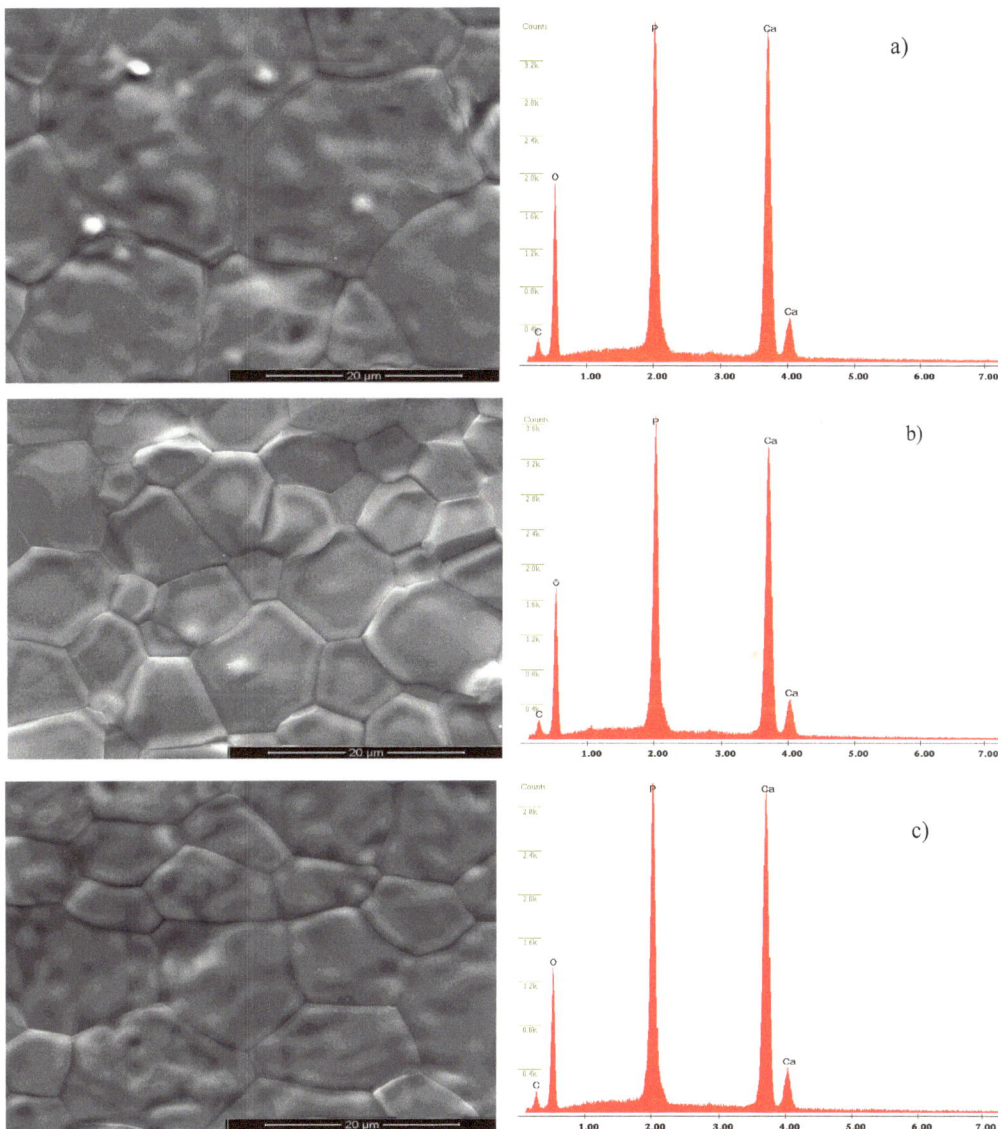

Figure 6: Representative SEM images and EDS spectra of the prepared composites: (a) Si-HA; (b) Ce-HA; and (c) La-HA.

Image analysis revealed a well-defined structure, composed by HA and phosphosilicate glass grains. The surface was continuous and crack-free, indicative of a strong bonding between the grains. In the analysis of the EDS spectra, doped glass elements were not detected, although these elements were identified by the same analytic procedure conducted on the previously prepared glasses. This might be related to the fact that the composites contain only a small amount of added glass (2.5%wt) and the system does not present enough resolution to detect their signal.

Analysis also revealed a Ca/P ratio of approximately 1.7-1.8, for all the evaluated samples. In pure and stoichiometric HA this ratio is 1.6. Thus, the prepared composites are made of more than one calcium-phosphate phase. These results were verified by XRD analysis and are presented in Fig. **7.**

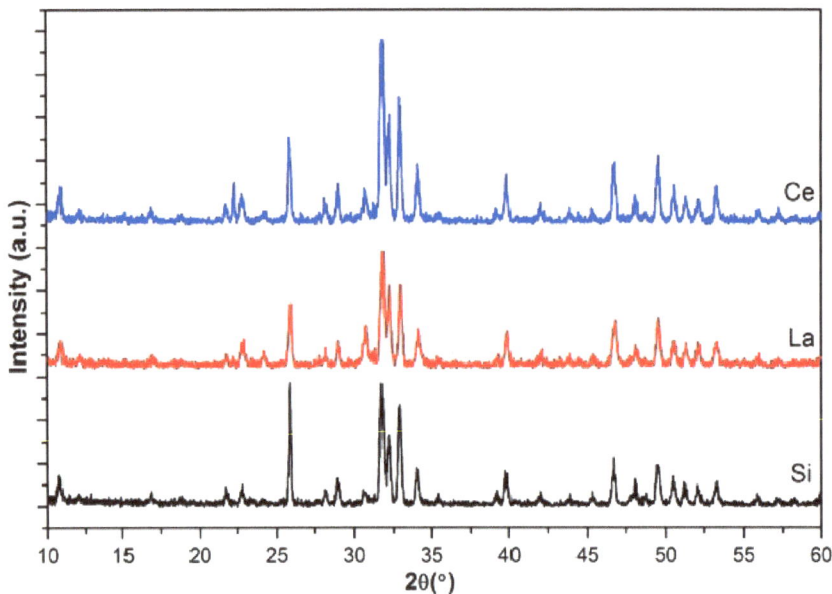

Figure 7: XRD spectra of the Si-HA, Ce-HA and La-HA composites.

In this figure, the HA peak is clearly observed at around 31.7°. However, the spectra also revealed some small differences between the samples. In order to clarify the origin of these variations, a chemical phase quantitative analysis was performed. The results are presented in Table **4.**

Table 4: Chemical phases present in the GR-HA composites (Si-HA, Ce-HA and La-HA)

Composites	HA	α- TCP	β -TCP	Others	
				Ca$_4$O(PO$_4$)$_2$	CaO
Si-HA	74.20	10.52	7.97	4.15	3.17
Ce-HA	64.62	20.16	8.11	4.21	2.90
La-HA	71.07	14.10	8.38	4.07	2.38

It is possible to observe the formation of new chemical phases, apart from HA - namely α-TCP and β-TCP, induced by the synthesis process, thus substantiating the variation in the Ca/P ratio of the materials obtained.

Calcium phosphate and its analogs constitute around 60–70% of the bone tissue and are known to contribute to their mechanical properties. HA is the most predominant form and is known to remain a stable calcium phosphate compound, even at low pH conditions. It is classed as bioactive, *i.e.* known to establish ion-exchange reactions with the surrounding body fluids, thus resulting in the formation of a biologically active carbonate apatite layer that establishes a strong chemical bond with the adjacent tissue [59-61]. On the other hand, TCP is a bioresorbable calcium phosphate compound, characterized by the ability to be replaced by newly formed ingrowth bone tissue. The resorbable TCP serves as a rich calcium and phosphorus source, which can be easily assimilated and absorbed. As a biomaterial, it has been found to be highly biocompatible and to create a resorbable interlocking network within the neighbouring tissue, in order to promote healing [59-61].

Biomaterials containing both HA and TCP phases seem to share a similar composition, structure and characteristics with the native human bone, expressing an improved biocompatibility when assessed, both *in vitro* and *in vivo*. HA/TCP biomaterials were shown to report an adequate mechanical fixation to the homing bone and to guide bone tissue ingrowth, facilitate bone formation and consolidation [62-64]. These composites have been successfully used in the clinics as bone substitutes.

GR-HA composite samples were biologically characterized by *in vitro* methodologies. Cell culture methods provide a necessary and useful adjunct to *in vivo* studies in the biological analysis of potential biomaterials [65]. One of the most important functions is the screening for toxic effects of the biomaterials. Cellular behaviour that ranges from cell death to alterations of cell adhesion, proliferation and biosynthetic activity can be observed [66, 67]. It is thus possible to construct a range of *in vitro* changes, ranging from marked inhibition of growth with well demarked cell death – that broadly describes the material as not biocompatible - to marked enhancement of cell proliferation and/or other cell biological parameters - describing the material as biocompatible and/or bioactive [65].

3.3. Biological Evaluation of GR-HA Composites with Human Osteoblastic-like Cells

In this study, human osteosarcoma-derived MG-63 osteoblastic-like cells were used. This human cell line has long been used for cytotoxicity and biocompatibility assays within the biological evaluation of biomaterials for bone tissue repair/regeneration. Briefly, at a low density, they report a characteristically oval to spindle-shaped morphology, with fine branching cytoplasmic cell processes, similar to those of human osteoblasts in culture [68, 69]. This cell line also expresses several characteristic markers of osteoblasts, that are positive for alkaline phosphatase, osteocalcin, osteonectin, osteopontin, osteoprotegerin, collagen type I, bone morphogenetic proteins, collagen type III, bone sialoprotein and decorin [68]. Also, MG-63 cells have been shown to express Cbfa-1/Runx-2, the earliest known transcription regulator of osteoblast differentiation and consequently demonstrated the commitment of these cells toward the osteogenic pathway [70].

In this experimental work, MG-63 human osteoblast-like cells were directly cultured on the surface of Si-HA, Ce-HA, La-HA composites and HA samples for 5 days. The assessment of the biological response of the seeded materials was conducted by the MTT assay and confocal laser scanning microscopy, following staining for F-actin cytoskeleton and nucleus counterstaining.

The results regarding cell viability/proliferation (MTT assay) are shown in Fig. **8**. MTT is metabolised to a purple formazan salt by mitochondrial enzymes in living cells and the absorbance is proportional to the number of viable cells in the sample. In Fig. **8**, the data reflects both the quantitative values of the MTT reduction assay (Fig. **8A**) and the qualitative image analysis of MTT staining (Fig. **8B**). Cells growing on the material samples presented a high proliferation rate throughout the entire culture period. At early incubation times (2 days), values of MTT reduction were significantly higher on the composites, comparing to control (HA samples). This is indicative of a higher amount of viable cells attached to seeded composites and where able to initiate cell division. Addressing the GR-HA composites, a significantly increased MTT reduction value was obtained within La-HA samples, compared to the other materials. This might substantiate an improved biological behaviour of this composition (at early times) by the specific enhancement of cell adhesion and promotion of early proliferative events.

Subsequently, cells growing over the surface of GR-HA composites presented an active proliferation. At 5 days of culture values for the MTT reduction were significantly higher than those of the control (seeded HA samples). Within the assessment of the GR-HA behaviour, La-HA and Ce-HA composites revealed a significantly higher value for the MTT reduction at day 5, compared to the non-doped GR-HA composite (Si-HA).

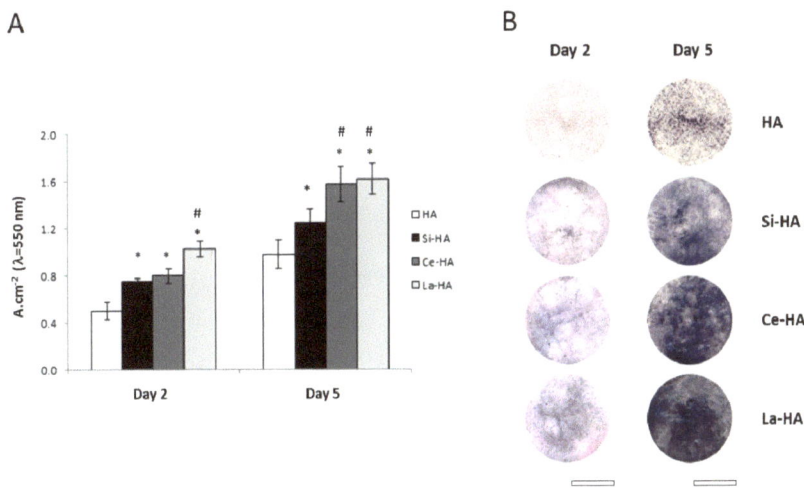

Figure 8: A - Cell viability/proliferation of MG63 cells cultured on GR-HA composites (Si-HA, Ce-HA and La-HA) and HA for 5 days, estimated by the MTT assay.* - significantly different from HA. # - significantly different from Si-HA. B –representative stereo microscope images of the surface of seeded samples at 2 and 5 days of culture. Scale bar represents 1 cm.

The time-course behaviour of seeded materials was also monitored using confocal laser scanning microscopy (Figs. **9** and **10**). Low magnification images revealed that at day 2 of culture (Fig. **9**) seeded composites were densely colonised with MG63 cells, showing areas of high cell density. Compared to control (HA)

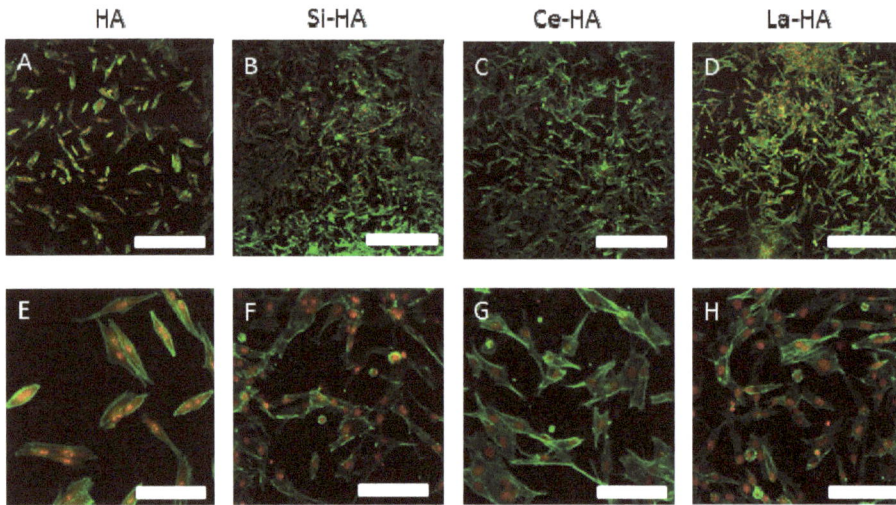

Figure 9: Confocal laser scanning microscopy micrographs of seeded MG-63 human osteoblastic-like cells, for 2 days, over HA (A, E), Si-HA (B, F), Ce-HA (C, G) and La-HA (D, H). Phalloidin (green) and propidium iodide (red) labelling. A to D, 100x magnification – scale bar corresponds to 300 μm; E to H, 400x magnification.

samples, seeded composites exhibited an increased adhesion and early proliferation of the seeded cells, with established cell-to-cell interconnectivity. Among seeded GR-HA composites, qualitatively, an increased cell density was found on the surface of La-HA samples. This was assessed by the presence of organised cell clusters, observed at day 2 of culture. High magnification images revealed that the composites (Si-HA, Ce-HA and La-HA) exhibited the normal morphological display of human osteoblastic-like cells. Compared to seeded HA (control), a similar degree of cytoplasmic spreading, as well as the setting up of prominent nuclear structures and the establishment of numerous thin cytoplasmatic projections responsible for the extensive intercellular contacts, were verified. The early presence of the *in vitro* osteoblastic morphology following the adhesion process is in agreement with the suitable reorganisation of the cytoskeleton during the first 24-48 hours of culture, and is thoroughly verifiable over the assayed composites. This behaviour is highly correlated with the absence of substrate-induced cytotoxic effects. The early cytoskeleton rearrangement plays a determinant role in the fate of several subsequent cellular events because of its ability to modulate intracellular trafficking of mediators. These may contribute to modulation of the cell morphology, the capacity for cell proliferation and differentiation over the seeded substrates [71, 72]. Therefore, the assessment of the early cytoskeleton organisation has been regarded as an important assay in the evaluation of the *in vitro* biological behaviour.

The cells proliferated actively throughout the culture time and, at day 5 (Fig. **10**), a densely organised cell layer could be seen for the control (HA) and seeded GR-HA composites. Low-magnification and high-magnification micrographs are presented. An increased proliferation – as compared to control – was achieved in the Si-HA, Ce-HA and La-HA composite surface, as shown by the establishment of organised cell multilayers, in contrast with the control surface, in which only an organised cell monolayer was seen.

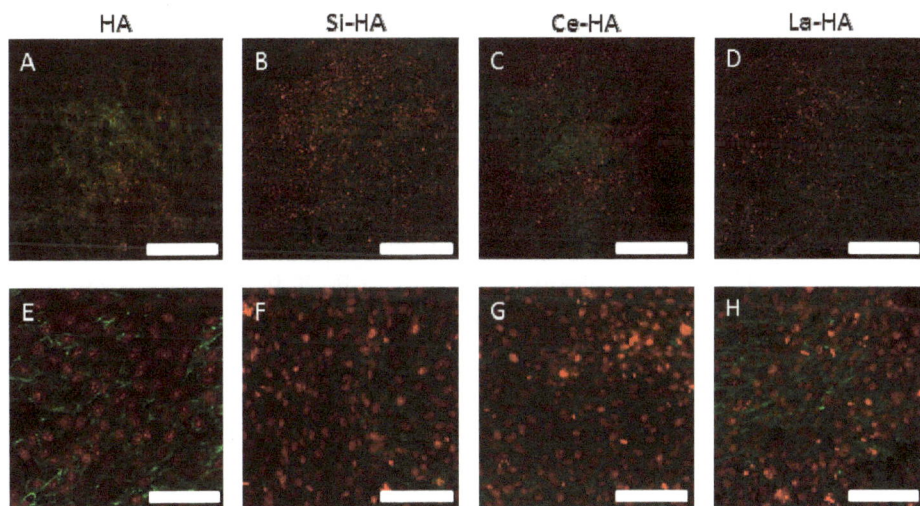

Figure 10: Confocal laser scanning microscopy micrographs of seeded MG-63 human osteoblastic-like cells, for 5 days, over HA (A, E), Si-HA (B, F), Ce-HA (C, G) and La-HA (D, H). Phalloidin (green) and propidium iodide (red) labeling. A to D, 100x magnification – scale bar corresponds to 300 μm; E to H, 400x magnification.

In the present work, GR-HA composites were developed with rare earth elements that allowed an active osteoblastic cell adhesion and proliferation throughout the culture period, as assayed by the MTT assay and confocal laser scanning microscopy. Comparatively to control (HA) samples, an increased cell adhesion and proliferation were obtained, especially in those composites containing a bioactive glass doped with RE elements.

Rare earth elements have been used within a wide range of biomedical applications [21-23]. More recently, their biological activity was found to modulate bone tissue metabolism and also the bone regeneration process. *In vivo* reports showed that, following oral administration for 6 months, La was found to accumulate in the bone tissue of male Wistar rats. Thermogravimetric analysis revealed a decrease in the mineral-to-matrix ratio and an increase in carbonate content, compared to a control. Further analysis revealed a smaller mean thickness of the mineral crystals and a lowered disorder in the crystals. Also, within these smaller sized crystals more adsorbed labile carbonate and more acidic phosphate made the bone mineral easier to dissolve, as revealed in the kinetic measurement of bone demineralization [27]. Interestingly, Behets *et al.* found that the administration of high doses of $La_2(CO_3)_3.4H_2O$ for 3 months made no difference to the bone histomorphometric analysis of rats with normal renal function [73]. In accordance, lanthanum carbonate octahydrate given orally at a dose of 1g/Kg for 3 months did not seem to alter the histological microstructure of the skeleton of healthy cats [74].

Experimental *in vitro* test were also conducted in order to assess the biological effects of specific lanthanides on the behaviour of bone cell populations. La^{3+} was shown to inhibit the osteoblastic differentiation of rat mesenchymal stem cells in the early and middle stages of culture, as demonstrated by the decrease in the alkaline phosphatase activity and osteocalcin expression. La^{3+} effects seemed to be dependent on the activation of the mitogen-activated protein kinase (MAPK) and not to effect the matrix mineralization process; interestingly the presence of La^{3+} on the culture microenvironment up-regulated the type I collagen and osteocalcin expression. It also enhanced the ALP activity at late culture periods [75]. Another report by Wang *et al.* showed that La^{3+} treatment enhanced the osteoblastic differentiation of calvaria-derived rat osteoblasts, as seen by the enhancement of ALP activity, osteocalcin expression and matrix mineralization. Furthermore, the expressions of osteoblast-specific genes (Cbfa-1, osteopontin, and bone sialoprotein) were increased, while no changes were observed regarding the expression of type I collagen. These results were shown to be at least, partially mediated by the enhanced phosphorylation of extracellular signal-regulated kinase (ERK) [32]. Ce^{3+} has been shown to promote the proliferation of mouse osteoblasts over a wide range of concentrations and to inhibit the adipocytic transdifferentiation of these cells. High concentrations were shown to induce the osteoblastic phenotype, as assessed by the increased presence of mineralized extracellular matrix nodules at late culture periods [33]. Also, the effects of Dy^{3+} were assessed on the *in vitro* behaviour of both mouse primary bone marrow stromal cells (BMSC) and mouse osteoblasts. Dy^{3+} was shown to promote the osteogenic differentiation and inhibit the adipogenic differentiation of BMSC over a wide range of concentrations. It was also shown to induce the osteogenic phenotype of cultured mouse osteoblasts, as assessed by the expression of alkaline phosphatase and extracellular matrix mineralization [76].

Literature reports converge to substantiate that lanthanides are able to modulate the behaviour of bone-relevant cellular populations. In general, their biological behaviour basically originates from their similarity to the calcium ion. This is of the utmost relevance, since calcium signals have been found to play a role in a wide range of cellular processes including exocytosis, contraction, cell proliferation, differentiation and apoptosis [77, 78]. Interestingly, rare earth ions also seem to directly interfere with the intracellular calcium ionic concentration in a bidirectional way: on one hand, they seem to be able to increase the intracellular calcium ion by activating a membrane-bound Ca^{2+}-sensing receptor; while, on the other hand, they may decrease intracellular calcium ion by blocking calcium-dependent channels [79]. This two-fold mechanism might help expain the differences in the literature reports, which may also relate to the wide range of experimental models used, culture conditions and evaluated parameters. Moreover, the different species of rare earth ions and the wide concentration range in which they have been assayed may also contribute to substantiate the observed differences in the biological effect of lanthanides. Furthermore, differences may also relate to the physico-chemical characteristics of the respective cations, depending upon features such as their ionic radii or charge densities [80].

Regarding the assessment of the biological properties of RE ions and their suitability for the enhancement of bone tissue regeneration, plus their use in the constitution of biomaterials targeting bone tissue repair/regeneration, this has been assessed. Metal alloys containing rare-earth elements have been developed with success. For instance, a free machining titanium alloy $Ti_6Al_4VLa_{0.9}$ was developed and shown to display adequate biomechanical properties, improved machinability and a cytocompatibility similar to the one of the reference alloy [81]. Rare earth elements have also been added to magnesium degrading metal alloys, within the development of bone implants with high primary stability [82]. Following long term intramedullary implantation in the femurs of guinea pigs, RE-containing alloys reported a corrosion layer accumulation in direct contact with the surrounding bone, with high mineral apposition rates and an increased bone mass around the implanted rod compared to a control [82]. Lanthanides have also been added to ceramic materials and shown to improve their biological response. La-doped HA has been found to enhance the adhesion and differentiation of human osteoblastic cells at earlier times than either HA doped with divalent cations (Mg^{2+} and Zn^{2+}) or undoped HA [36]. Guo *et al.* highlighted that La incorporation in HA gave an increase in flexural strength, a lower dissolution rate, higher thermal properties and excellent biocompatibility, in comparison to pure phase HA [38]. Furthermore, the nanotoxicity assessment of HA nanoparticles co-doped with Eu^{3+} and Gd^{3+} carried out on primary human endothelial cells, a mouse fibroblast cell line, human nasopharyngeal carcinoma cells and on a human lung cancer cell line revealed no apparent toxicity, up to the relatively high doses of 500 µg/mL [39].

These results are indicative of the absence of cytotoxicity of bone-related biomaterials containing RE elements and support the trend for the enhancement of the biological behaviour of rare earth containing biomaterials, in comparison to the reference substrates. This is in line with data from this experimental study, in which La-HA and Ce-HA composites revealed an improved biological behaviour, by allowing an enhanced human osteoblastic-like cell adhesion and proliferation, compared to a control (HA) and to non-doped composites.

CONCLUSION

This work reports the synthesis of the glass system SiO_2-P_2O_5 doped with cerium and lanthanum oxides. By means of FTIR and Raman spectroscopy a normal structure for the glasses of the SiO_2-P_2O_5 system was observed, while the addition of dopants ions (lanthanides – Ce and La) induced a slight change the internal network. The morphological characterisation, performed by SEM, revealed smooth and glassy surfaces for all the assayed samples. Furthermore, the EDS spectra revealed the expected composition with no traces of contamination. The glasses were used to prepare the glass-HA composites, composed by 97.5 wt% HA and 2.5 wt% glass, with biomedical applications within bone tissue regeneration in mind. The composites obtained were characterised by SEM, EDS and XRD, exhibited, apart from preponderance of HA, mostly α, β-TCP phases, and also remnants of $Ca_4O(PO_4)_2$ and CaO. Preliminary biological characterisation, conducted with human osteoblastic-like cells, revealed an improved biological behaviour of doped composites, which allowed an improved cellular adhesion and proliferation throughout the culture time, in comparison to HA and a non-doped glass-HA composite.

CONFLICT OF INTEREST

The contents presented in the chapter have been carefully written based on the review from the references cited and the results obtained from the investigations carried out by the authors. Further, there is no conflict of interest with other people or organisations in respect of the present research work.

ACKNOWLEDGMENTS

The authors would like to thank FCT- *Fundacão para e Ciencia Tecnologia* (Ref. No. PTDC/SAU-BEB/103034/2008) due to their financial support.

REFERENCES

[1]   Vallet-Regí M. Evolution of bioceramics within the field of biomaterials. Comptes Rendus Chimie 2010;13(1):174-85.

[2]   Ravarian R, Moztarzadeh F, Hashjin MS, Rabiee SM, Khoshakhlagh P, Tahriri M. Synthesis, characterization and bioactivity investigation of bioglass/hydroxyapatite composite. Ceram Int 2010;36(1):291-7.

[3]   Doostmohammadi A, Monshi A, Fathi MH, Braissant O. A comparative physico-chemical study of bioactive glass and bone-derived hydroxyapatite. Ceram Int 2011;37(5):1601-7.

[4]   Veljović D, Jančić-Hajneman R, Balać I, Jokić B, Putić S, Petrović R, *et al*. The effect of the shape and size of the pores on the mechanical properties of porous HAP-based bioceramics. Ceram Int 2011;37(2):471-9.

[5]   Sprio S, Tampieri A, Celotti G, Landi E. Development of hydroxyapatite/calcium silicate composites addressed to the design of load-bearing bone scaffolds. J Mech Behav Biomed Mater 2009;2(2):147-55.

[6]   Szpiro L, Park JB, Kumar NM. Bioceramics: properties, characterizations, and applications: Springer; 2008.

[7]   Macaskie LE, Yong P, Paterson-Beedle M, Thackray AC, Marquis PM, Sammons RL, *et al.* A novel non line-of-sight method for coating hydroxyapatite onto the surfaces of support materials by biomineralization. J Biotechnol 2005;118(2):187-200.

[8]   Hu Y, Miao X. Comparison of hydroxyapatite ceramics and hydroxyapatite/borosilicate glass composites prepared by slip casting. Ceram Int 2004;30(7):1787-91.

[9]   Knowles JC, Talal S, Santos JD. Sintering effects in a glass reinforced hydroxyapatite. Biomaterials 1996;17(14):1437-42.

[10]   Lopes MA, Monteiro FJ, Santos JD. Glass-reinforced hydroxyapatite composites: fracture toughness and hardness dependence on microstructural characteristics. Biomaterials 1999;20(21):2085-90.

[11]   Queiroz A, Santos J, Monteiro F, Prado da Silva M. Dissolution studies of hydroxyapatite and glass-reinforced hydroxyapatite ceramics. Mater Charact 2003;50(2-3):197-202.

[12]   Kokubo T, Materials JM. Bioceramics and their clinical applications: Woodhead Pub. and Maney Pub. on behalf of Institute of Materials, Minerals & Mining 2008.

[13]   Santos JD, Jha LJ, Monteiro FJ. Surface modifications of glass-reinforced hydroxyapatite composites. Biomaterials 1995;16(7):521-6.

[14]   Kannan S, Vieira SI, Olhero SM, Torres PMC, Pina S, da Cruz e Silva OAB, *et al.* Synthesis, mechanical and biological characterization of ionic doped carbonated hydroxyapatite/β-tricalcium phosphate mixtures. Acta Biomater 2011;7(4):1835-43.

[15]   Laurencin D, Almora-Barrios N, de Leeuw NH, Gervais C, Bonhomme C, Mauri F, *et al.* Magnesium incorporation into hydroxyapatite. Biomaterials 2011;32(7):1826-37.

[16]   Miao S, Lin N, Cheng K, Yang D, Huang X, Han G, *et al.* Zn-Releasing FHA Coating and Its Enhanced Osseointegration Ability. J Am Ceram Soc 2011;94(1):255-60.

[17]   Hoppe A, Güldal NS, Boccaccini AR. A review of the biological response to ionic dissolution products from bioactive glasses and glass-ceramics. Biomaterials 2011;32(11):2757-74.

[18]   Aminian A, Solati-Hashjin M, Samadikuchaksaraei A, Bakhshi F, Gorjipour F, Farzadi A, *et al.* Synthesis of silicon-substituted hydroxyapatite by a hydrothermal method with two different phosphorous sources. Ceram Int 2011;37(4):1219-29.

[19]   Lin Y, Yang Z, Cheng J. Preparation, Characterization and Antibacterial Property of Cerium Substituted Hydroxyapatite Nanoparticles. J Rare Earths 2007;25(4):452-6.

[20]   Aissa A, Debbabi M, Gruselle M, Thouvenot R, Flambard A, Gredin P, *et al.* Sorption of tartrate ions to lanthanum (III)-modified calcium fluor- and hydroxyapatite. J Colloid Interface Sci 2009;330(1):20-8.

[21]   Fricker SP. The therapeutic application of lanthanides. Chem Soc Rev 2006;35(6):524-33.

[22]   Thompson KH, Orvig C. Editorial: Lanthanide compounds for therapeutic and diagnostic applications. Chem Soc Rev 2006;35(6):499-.

[23]   Matsuda T, Yamanaka C, Ikeya M. ESR study of Gd^{3+} and Mn^{2+} ions sorbed on hydroxyapatite. Appl Radiat Isot 2005;62(2):353-7.

[24]   Fan Y, Huang S, Jiang J, Li G, Yang P, Lian H, *et al.* Luminescent, mesoporous, and bioactive europium-doped calcium silicate (MCS: Eu^{3+}) as a drug carrier. J Colloid Interface Sci 2011;357(2):280-5.

[25]   Neumeier M, Hails LA, Davis SA, Mann S, Epple M. Synthesis of fluorescent core-shell hydroxyapatite nanoparticles. J Mater Chem 2011; 21(4):1250-4.

[26]   Appavoo IA, Zhang Y. Upconverting Fluorescent Nanoparticles for Biological Applications. Emerging Nanotechnologies for Manufacturing 2010:159-75.

[27]   Huang J, Zhang TL, Xu SJ, Li RC, Wang K, Zhang J, *et al.* Effects of Lanthanum on Composition, Crystal Size, and Lattice Structure of Femur Bone Mineral of Wistar Rats. Calcif Tissue Int 2006;78(4):241-7.

[28]   Barta CA, Sachs-Barrable K, Jia J, Thompson KH, Wasan KM, Orvig C. Lanthanide containing compounds for therapeutic care in bone resorption disorders. Dalton Transactions 2007(43):5019-30.

[29]   Herath HMTU, Di Silvio L, Evans JRG. *In vitro* evaluation of samarium (III) oxide as a bone substituting material. J Biomed Mater Res A 2010;94A(1):130-6.

[30]   Quarles LD, Hartle JE, Middleton JP, Zhang J, Arthur JM, Raymond JR. Aluminum-Induced DNA synthesis in osteobalsts: Mediation by a G-protein coupled cation sensing mechanism. J Cell Biochem 1994;56(1):106-17.

[31]   Dawei Z, Jinchao Z, Yao C, Mengsu Y, Xinsheng Y. Effects of lanthanum and gadolinium on proliferation and differentiation of primary osteoblasts. Prog Nat Sci 2007;17(5):618-23.

[32]   Wang X, Yuan L, Huang J, Zhang T-L, Wang K. Lanthanum enhances *in vitro* osteoblast differentiation *via* pertussis toxin-sensitive gi protein and ERK signaling pathway. J Cell Biochem 2008;105(5):1307-15.

[33]   Zhang J, Liu C, Li Y, Sun J, Wang P, Di K, *et al.* Effect of cerium ion on the proliferation, differentiation and mineralization function of primary mouse osteoblasts *in vitro.* J Rare Earths 2010;28(1):138-42.

[34]   Zhang J, Zhang T, Xu S, Wang K, Yu S, Yang M. Effects of lanthanum on formation and bone-resorbing activity of osteoclast-like cells. J Rare Earths 2004;22(6):891.

[35]   Zhang J, Xu S, Wang K, Yu S. Effects of the rare earth ions on bone resorbing function of rabbit mature osteoclasts*in vitro.* Chin Sci Bull 2003;48(20):2170-5.

[36]   Webster TJ, Massa-Schlueter EA, Smith JL, Slamovich EB. Osteoblast response to hydroxyapatite doped with divalent and trivalent cations. Biomaterials 2004;25(11):2111-21.

[37]   Ahymah Joshy MI, Elayaraja K, Suganthi RV, Chandra Veerla S, Kalkura SN. *In vitro* sustained release of amoxicillin from lanthanum hydroxyapatite nano rods. Curr Appl Phys 2011;11(4):1100-6.

[38]   Guo DG, Wang AH, Han Y, Xu KW. Characterization, physicochemical properties and biocompatibility of La-incorporated apatites. Acta Biomater 2009;5(9):3512-23.

[39]   Ashokan A, Menon D, Nair S, Koyakutty M. A molecular receptor targeted, hydroxyapatite nanocrystal based multi-modal contrast agent. Biomaterials 2010;31(9):2606-16.

[40]   Ahmed I, Lewis M, Olsen I, Knowles J. Phosphate glasses for tissue engineering: Part 1. Processing and characterisation of a ternary-based P_2O_5-CaO-Na_2O glass system. Biomaterials 2004;25(3):491-9.

[41]   Sitarz M. Influence of modifying cations on the structure and texture of silicate–phosphate glasses. J Mol Struct 2008;887(1):237-48.

[42]   Doweidar H. Density-structure correlations in Na_2O-CaO-P_2O_5-SiO_2 bioactive glasses. J Non-Cryst Solids 2009;355(9):577-80.

[43]   Santos R, Santos LF, Almeida RM. Optical and spectroscopic properties of Er-doped niobium germanosilicate glasses and glass ceramics. J Non-Cryst Solids 2010;356(44-49):2677-82.

[44]   Shih P. Properties and FTIR spectra of lead phosphate glasses for nuclear waste immobilization. Mater Chem Phys 2003;80(1):299-304.

[45]   Stoch L, Sroda M. Infrared spectroscopy in the investigation of oxide glasses structure. J Mol Struct 1999;511:77-84.

[46]   Aguiar H, Serra J, González P, León B. Structural study of sol–gel silicate glasses by IR and Raman spectroscopies. J Non-Cryst Solids 2009;355(8):475-80.

[47]   Szumera M, Waclawska I, Mozgawa W, Sitarz M. Spectroscopic study of biologically active glasses. J Mol Struct 2005;744:609-14.

[48]   Ming C, Greish Y, El-Ghannam1 A. Crystallization behavior of silica-calcium phosphate biocomposites: XRD and FTIR studies. J Mater Sci Mater Med 2004;15(11):1227-35.

[49]   Nieminen M, Putkonen M, Niinistö L. Formation and stability of lanthanum oxide thin films deposited from [beta]-diketonate precursor. Appl Surf Sci 2001;174(2):155-66.

[50]   Ansari AA, Kaushik A, Solanki P, Malhotra B. Sol-gel derived nanoporous cerium oxide film for application to cholesterol biosensor. Electrochem Commun 2008;10(9):1246-9.

[51]   Román J, Padilla S, Vallet-Regí M. Sol–Gel Glasses as Precursors of Bioactive Glass Ceramics. Chem Mater 2003;15(3):798-806.

[52]   Roiland C, Fayon F, Simon P, Massiot D. Characterization of the disordered phosphate network in CaO–P2O5 glasses by 31P solid-state NMR and Raman spectroscopies. J Non-Cryst Solids 2011;357(7):1636-46.

[53]   Hudgens JJ, Brow RK, Tallant DR, Martin SW. Raman spectroscopy study of the structure of lithium and sodium ultraphosphate glasses. J Non-Cryst Solids 1998;223(1–2):21-31.

[54]   Vedeanu N, Cozar O, Ardelean I, Lendl B, Magdas DA. Raman spectroscopic study of CuO–V2O5–P2O5–CaO glass system. Vib Spectrosc 2008;48(2):259-62.

[55]   Hwa L-G, Shiau J-G, Szu S-P. Polarized Raman scattering in lanthanum gallogermanate glasses. J Non-Cryst Solids 1999;249(1):55-61.

[56]   Zhang F, Chan SW, Spanier JE, Apak E, Jin Q, Robinson RD, *et al.* Cerium oxide nanoparticles: Size-selective formation and structure analysis. Appl Phys Lett 2002;80(1):127-9.

[57]   Aguiar H, Solla EL, Serra J, González P, León B, Almeida N, *et al.* Orthophosphate nanostructures in SiO2–P2O5–CaO–Na2O–MgO bioactive glasses. J Non-Cryst Solids 2008;354(34):4075-80.

[58]   Uy D, O'Neill AE, Xu L, Weber WH, McCabe RW. Observation of cerium phosphate in aged automotive catalysts using Raman spectroscopy. Appl Catal, B 2003;41(3):269-78.

[59]   LeGeros RZ. Properties of osteoconductive biomaterials: calcium phosphates. Clin Orthop 2002;395:81.

[60]   Phang M, Angela M, Fuzina H, Aminuddin B, Ruszymah B, Fauziah O. Looking into the microstructure of biomaterials and its inhabitants in engineered bone tissue. J Microsc 2004;1:73–9.

[61]   Hutmacher DW, Schantz JT, Lam CXF, Tan KC, Lim TC. State of the art and future directions of scaffold-based bone engineering from a biomaterials perspective. J Tissue Eng Regen M 2007;1(4):245-60.

[62]   Wang Y, Ni M, Tang P-F, Li G. Novel application of HA-TCP biomaterials in distraction osteogenesis shortened the lengthening time and promoted bone consolidation. J Orthop Res 2009;27(4):477-82.

[63]   Ng AMH, Tan KK, Phang MY, Aziyati O, Tan GH, Isa MR, *et al.* Differential osteogenic activity of osteoprogenitor cells on HA and TCP/HA scaffold of tissue engineered bone. J Biomed Mater Res A 2008;85A(2):301-12.

[64]   G D. Biphasic calcium phosphate concept applied to artificial bone, implant coating and injectable bone substitute. Biomaterials 1998;19(16):1473-8.

[65]   Kirkpatrick C, Mittermayer C. Theoretical and practical aspects of testing potential biomaterials *in vitro*. J Mater Sci - Mater Med 1990;1(1):9-13.

[66]   Coelho MJ, Trigo Cabral A, Fernandes MH. Human bone cell cultures in biocompatibility testing. Part I: osteoblastic differentiation of serially passaged human bone marrow cells cultured in α-MEM and in DMEM. Biomaterials 2000;21(11):1087-94.

[67]   Coelho MJ, Fernandes MH. Human bone cell cultures in biocompatibility testing. Part II: effect of ascorbic acid, β-glycerophosphate and dexamethasone on osteoblastic differentiation. Biomaterials 2000;21(11):1095-102.

[68]   Pautke C, Schieker M, Tischer T, Kolk A, Neth P, Mutschler W, *et al.* Characterization of osteosarcoma cell lines MG-63, Saos-2 and U-2 OS in comparison to human osteoblasts. Anticancer Res 2004;24(6):3743.

[69]   Amaral M, Dias A, Gomes P, Lopes M, Silva R, Santos J, *et al.* Nanocrystalline diamond: *In vitro* biocompatibility assessment by MG63 and human bone marrow cells cultures. J Biomed Mater Res A 2008;87:91-9.

[70]   Yang S, Wei D, Wang D, Phimphilai M, Krebsbach PH, Franceschi RT. *In Vitro* and *In Vivo* Synergistic Interactions Between the Runx2/Cbfa1 Transcription Factor and Bone Morphogenetic Protein-2 in Stimulating Osteoblast Differentiation. J Bone Miner Res 2003;18(4):705-15.

[71]   K A. Osteoblast adhesion on biomaterials. Biomaterials 2000;21(7):667-81.

[72]   Wang N, Butler JP, Ingber DE. Mechanotransduction across the cell surface and through the cytoskeleton. Science 1993;260(5111):1124.

[73]   Behets GJ, Dams G, Vercauteren SR, Damment SJ, Bouillon R, De Broe ME, *et al.* Does the phosphate binder lanthanum carbonate affect bone in rats with chronic renal failure? J Am Soc Nephrol 2004;15(8):2219-28.

[74]   Nunamaker E, Sherman J. Oral administration of lanthanum dioxycarbonate does not alter bone morphology of normal cats. J Vet Pharmacol Ther 2011.

[75]   Liu H, Zhang T, Xu S, Wang K. Effect of La^{3+} on osteoblastic differentiation of rat bone marrow stromal cells. Chin Sci Bull 2006;51(1):31-7.

[76]   Zhang J, Liu D, Sun J, Zhang D, Shen S, Yang M. Effect of Dy^{3+} on osteogenic and adipogenic differentiation of mouse primary bone marrow stromal cells and adipocytic trans-differentiation of mouse primary osteoblasts. Chinese Scientific Bulletin 2009;54:66-71.

[77]   Michael J B. Calcium signal transduction and cellular control mechanisms. BBA-Mol Cell Res 2004;1742(1–3):3-7.

[78]   David E C. Calcium Signaling. Cell 2007;131(6):1047-58.

[79]   Riccardi D, Finney B, Wilkinson W, Kemp P. Novel regulatory aspects of the extracellular Ca^{2+}-sensing receptor, CaR. Pflugers Archiv European Journal of Physiology 2009;458(6):1007-22.

[80]   Wang K, Li R, Cheng Y, Zhu B. Lanthanides—the future drugs? Coord Chem Rev 1999;190–192(0):297-308.

[81]   Feyerabend F, Siemers C, Willumeit R, Rösler J. Cytocompatibility of a free machining titanium alloy containing lanthanum. J Biomed Mater Res A 2009;90A(3):931-9.

[82]   Witte F, Kaese V, Haferkamp H, Switzer E, Meyer-Lindenberg A, Wirth CJ, *et al. In vivo* corrosion of four magnesium alloys and the associated bone response. Biomaterials 2005;26(17):3557-63.

CHAPTER 5

Calcium Phosphate Ceramics in Periodontal Regeneration

Pavan K. Gudi[1], Sooraj H. Nandyala[2,3,*], Pedro. S. Gomes[4], Maria A. Lopes[5], Maria H. Fernandes[4] and José D. Santos[5]

[1]*Department of Periodontics, Govt. Dental College and Hospital, Hyderabad – 500 012 AP, India;* [2]*INESC Porto, Rua do Campo Alegre, 687, 4169-007 Porto, Portugal;* [3]*Departamento de Física, Faculdade de Ciências, Universidade do Porto, Rua do Campo Alegre, 687, 4169-007 Porto, Portugal;* [4]*Laboratório de Farmacologia e Biocompatibilidade Celular. Faculdade de Medicina Dentária, Universidade do Porto, Rua Dr. Manuel Pereira da Silva, 4200-393 Porto, Portugal and* [5]*DEMM, Faculty of Engineering, University of Porto, Rua Dr. Roberto Frias, 4200-465 Porto, Portugal*

Abstract: Regenerative periodontal therapy aims to predictably restore the tooth's supporting periodontal tissues and should result in the formation of a new connective tissue attachment (*i.e.* new cementum with inserting periodontal ligament fibres) and new alveolar bone. This chapter aims to address the clinical application of bone grafts on periodontal regenerative approaches, with given relevance to the use of calcium phosphate ceramics. Furthermore, a clinical case is presented in which the regenerative capability of a glass-reinforced hydroxyapatite (Bonelike®) is thoroughly evaluated by clinical and tomographic measurements in the healing of a periodontal intrabony defect.

Keywords: Calcium phosphate materials, intrabony periodontal defects, periodontal regeneration, autografts, allografts, alloplasts, Bonelike®, bone graft, bioactive glass,bone regeneration, bone fill, calcium phosphate materials, hydroxyapatite, intrabony defects, open flap debridement, periodontal regeneration, probing depth, periodontal defect, periodontal surgery, tricalcium phosphate.

1. PERIODONTAL REGENERATION

1.1. Periodontal Disease

Periodontal disease is an inflammatory condition, inherited or acquired, that affects the surrounding and supporting tissues of the teeth. It is characterized by a chronic oral bacterial infection that results in inflammation of the gums, which thus leads to the gradual destruction of periodontal tissues, including the alveolar

*Address correspondence to Sooraj H. Nandyala: FCUP/INESC Porto/Department of Physics, Faculty of Sciences, University of Porto, Rua do Campo Alegre, 687, 4169-007 Porto, Portugal; Tel: +351 22 0402302; Fax: +351 22 0402 437; E-mail: nandyala.sooraj@fc.up.pt

Sooraj H. Nandyala and José D. Santos (Eds)
All rights reserved-© 2013 Bentham Science Publishers

bone that embraces teeth support [1]. Epidemiologic data estimates differ on the basis of the definition of the disease, its prevalence, severity, and rate of disease progression, and clearly fluctuates worldwide by race and geographic area, with older populations typically experiencing higher rates of periodontitis [2]. Estimates of the global prevalence of severe periodontal disease are broadly around 10-15%, although up to 90% of the population may be affected by some form of milder periodontal disease including gingivitis [2, 3].

In addition to age and race, other known risk factors for periodontal disease include gender, body mass index (obesity), tobacco consumption, stress and nutrition, as well as systemic conditions like diabetes, osteoporosis and neutrophilic dysfunctions [4-6]. Populations with limited access to dental healthcare and also those with a low socio-economic status may also be at an increased risk for the establishment and development of periodontal disease [5]. Furthermore, data of genetic polymorphisms associated with periodontal disease have been reported, noticeably those related to the interleukine-1 gene cluster [7].

Periodontal disease is a pathology with infectious etiology, caused broadly by anaerobic Gram-negative bacteria. Socransky *et al.* examined over 13,000 sub-gingival plaque samples from 185 adult subjects and used cluster analysis and community ordination techniques to group the identified bacterial species. Six closely associated clusters were consistently recognized and subsequently color-coded into their respective complexes, *i.e.* "Blue", "Green", "Yellow", "Purple", "Orange" and "Red". The first 4 complexes were described to be early colonizers of the tooth surface and to form the conditioning biofilm before the proliferation of the more pathogenic "Orange" and "Red" complexes [8]. The "Red" complex was shown to be strongly related to pocket depth and bleeding on probing – clinical signals of periodontal disease. Further, it has been shown that during the biofilm maturation, organisms from the "Orange" complex are required for the further establishment and colonization of the "Red" complex. The presence of these two complexes, in particular the "Red" complex, has been shown to be strongly correlated with severe and advanced stages of periodontal disease. In the "Red" complex, *P. gingivalis*, *T. denticola* and *T. forsythia* are included, while the "Orange" complex includes *F. nucleatum/periodonticum* subspecies*, P. intermedia, Prevotella nigrescens, P. micros, C. rectus, Campylobacter gracilis,*

Campylobacter showae, Eubacterium nodatum and *Streptococcus constellatus* [8]. Interestingly, many of these bacteria also seem to be present at low levels in the dental plaque of healthy individuals [9]. Overall, periodontal disease evolution seems to be signaled by a shift in the makeup of the dental biofilm from largely aerobic Gram-positive bacteria to a pathogenic infectious state dominated by anaerobic Gram-negative organisms [9]. Thus, the onset of periodontal disease is not marked by the establishment of novel infectious strains, but rather by a shift in the dominant strains composing the dental plaque biofilm [10]. These microorganisms possess an array of virulence factors that enhance their infectivity and provide the ability to the organisms to multiply and persist in the periodontium [10]. While the etiology of periodontitis seems to be bacterial, it is becoming clear that the pathogenesis of the disease is mediated by the development of a coordinated host response that induces an inflammatory reaction, which is thus destructive to the periodontal tissues [11]. Initially, protective aspects of the host response include recruitment of specific cellular immune-relevant populations, production of protective antibodies, and possibly the release of anti-inflammatory cytokines (*e.g.* transforming growth factor-β (TGF-β), interleukin-4 (IL-4), IL-10, and IL- 12). Nonetheless, perpetuation of the host response due to a persistent bacterial challenge disrupts homeostatic mechanisms of control and results in the release of mediators including pro-inflammatory cytokines (*e.g.* IL-1, IL-6, tumor necrosis factor-α [TNF-α]), proteases (*e.g.* matrix metalloproteinases), and prostanoids (*e.g.* prostaglandin E2 [PGE2]) which can promote extracellular matrix destruction in the gingiva, periodontal ligament and stimulate bone resorption [12].

Clinically, periodontitis results in the formation of soft tissue pockets or deepened crevices between the gingiva and tooth root. Loss of the periodontal ligament and disruption of its attachment to the cementum, as well as resorption of alveolar bone, also occur. Together with loss of attachment, it attains an apical migration of the epithelial attachment along the root surface and the resorption of neighboring bone [13,14].

Treatment of periodontitis on a first base is directed to the control of the cause (causal therapy), *i.e.* reduction of the bacterial load on the tooth surface and sulcus, by mechanical treatment above and below the gum level (scaling and root planing, respectively), pharmacologic treatment (anti-inflammatory and antibacterial local and/or systemic approaches) when indicated, and instructing and motivating the patient for reinforced home oral hygiene techniques [15, 16]. These conventional nonsurgical therapeutic approaches and surgical periodontal flap procedures are generally successful in halting the progression of the disease but broadly result in soft tissue recession - associated with poor aesthetics and result in residual pockets formation. Further, which are thus difficult to clean effectively and lessen (affect) long-term prognosis. These limited outcomes can be circumvented or underrated by periodontal regenerative procedures that aim to restore the lost function and anatomic organization of the periodontal structures.

1.2. Bone Grafts in Periodontal Regenerative Approaches

The goal of periodontal regeneration is the restoration of the periodontium to its original anatomical form and biological function. Regenerative periodontal therapy aims to predictably restore the tooth's supporting periodontal tissues and should result in formation of a new connective tissue attachment (*i.e.* new cementum with inserting periodontal ligament fibers) and restoration of the lost alveolar bone level [17]. Clinical outcome parameters consistent with successful regenerative therapy include - reduced probing depth, increased clinical attachment level and radiographic evidence of bone fill.

Bone replacement grafts, including autogenous grafts from intraoral or extraoral sites, allografts, xenografts, and alloplastic bone substitutes are the most widely used treatment modalities for the regeneration of periodontal osseous defects. They seem to provide a structural framework for clot development, maturation, and remodeling in addition to initiating the biological processes that support bone formation in the established osseous defects. Bone grafting materials also exhibit a variable capacity to promote the coordinated formation of bone, cementum, and periodontal ligament when placed and retained in periodontal defects [18].

Bone grafts can play an important role in the correction of the osseous aspects of periodontal defects, either by the process of osteogenesis, osteoinduction or by osteoconduction [19]. An osteogenic material, such as cancellous bone/bone marrow, contains living cells that are capable of differentiation and formation of new bone tissue. Alternatively, osteoinductive materials can induce bone formation by recruiting undifferentiated mesenchymal cells and guide them into the osteogenic pathway, whereas osteoconductive materials act principally as an inert scaffolding support for new bone formation [19, 20].

Various systems have been used to classify bone replacement grafts. Generally, these are sorted according to its source, chemical composition and/or physical properties. However, current advances in material and biological sciences have increasingly blurred the existing boundaries between classes. Here, we focus on a classification system based on the origin of the graft, categorizing them into autografts, allografts, xenografts and alloplastic materials. Historically, autografts were the first bone replacement grafts to be reported for periodontal applications, while allogenic freeze-dried bone was introduced to periodontics in the early 1970's. Demineralized allogenic freeze-dried bone gained wider application in the late 1980's while xenografts and alloplastic materials application for periodontal use occurred around the same time [21].

1.2.1. Autografts

For many years, autografts have been considered the gold standard for bone regenerative clinical applications. The decision to use autogenous grafts necessitates consideration of the donor site, procurement technique and handling or processing of the harvested material. Autogenous bone can be harvested intraorally (*e.g.* from the chin area, tuber region, healed extraction sites and toothless jaw segments), with or without processing, to yield graft materials of different forms (*e.g.* cortical chips, osseous coagulum and bone blend) [22]. Alternatively, extraoral sites (*e.g.* iliac crests, ribs, cranium and tibial metaphyses) can be used [23]. Autogenous grafts are nonimmunogenic and contain osteoblasts and osteoprogenitor stem cells, which are, theoretically, capable of proliferating and differentiating into the osteogenic lineage.

Autogenous bone grafts have been applied to periodontal defects. Histologic findings from early case reports, as well as subsequent radiographic and clinical data attained from large case series and controlled clinical trials substantiate the potential use of autogenous bone/bone marrow grafts (from either introal and extraoral origins) to support periodontal regeneration in humans [19,24]. Nonetheless, one should consider that there are limitations to obtaining autogenous grafts, such as insufficient oral sites for harvesting, the requirement for a second surgical site and expected morbidity at the donor site. Moreover, autograft harvesting is associated with an estimated 8.5–20% of complications, mostly including the possibility of hematoma formation, blood loss, nerve injury, hernia formation, infection, arterial injury, fracture, cosmetic defects, tumor transplantation, and sometimes chronic pain at the donor site [22, 25, 26].

1.2.2. *Allografts*

Allograft materials (*i.e.* bone graft tissues transplanted between different individuals from the same species) have been used with success in the regeneration of the bone tissue. These allografts include frozen grafts, freeze-dried bone allografts and demineralized freeze-dried bone allografts [18]. The possibility of disease transfer, antigenicity and the need for extensive cross-matching has precluded the use of fresh frozen bone in current clinical applications. Additionally, the evidence that freeze-drying markedly reduces the antigenicity and other health risks associated with fresh frozen bone, as well as the favorable results obtained in the field trials with freeze-dried bone allografts have led to the extensive use of these grafts in the treatment of periodontal osseous defects [27]. The use of cancellous bone, rather than cortical bone is broadly recommended since cancellous allografts were proven to be less antigenic and also due to the increased proportion of bone matrix, and therefore the presence of more osteoinductive components in the cancellous bone [28].

Further, research outputted that the bone matrix was responsible for blocking the biological availability of existing stimulating factors and thus, demineralized allografts were developed. Demineralized allogenic bone exhibits the capacity to induce bone formation in nonorthotopic sites, and is thus considered to be osteoinductive in nature. Bone demineralized to levels up to 2% residual calcium has been shown to provide maximum osteoinductive potential [29], presumably due to exposure of bone morphogenetic proteins [30]. Several works provided conclusive histologic evidence that demineralized allografts support periodontal regeneration in humans [18, 31].

The main advantages of using allografts are associated with the wide availability of the material and the absence of any donor site within the patient. Main reported disadvantages include the process of preparing the graft (*i.e.* freeze-drying and/or irradiating), which thus decrease the material's integrity and osteogenic potential, and the eventual immunogenicity of the material. Further, a major concern with allografts is the potential for disease transfer, particularly those associated with viral and prion pathogens [18, 19, 27].

1.2.3. Xenografts

Bone xenografts are naturally derived deproteinized cancellous bone grafts from another species, such as bovine, porcine, equine and natural coral. Following the withdrawal of the organic component, the enduring inorganic structure provides an organized matrix for new tissue ingrowth, thus maintaining the physical dimensions of the defect throughout bone regeneration [32]. However, there are still considerations regarding the immunological tolerance of the grafted tissues, the residual infection risk and a limited patient acceptance.

Mammals-derived bone xenografts present porosity and mineral content comparable to that of the human bone and perform as osteoconductive scaffolds, guiding the new bone tissue formation. These xenografts are prepared by chemical or low-heat extraction of the organic component from animal's bone. Histological evidence of the bone regeneration in periodontal defects has been attained with the use of bovine xenografts [33].

The calcium carbonate exoskeleton of coral species can be processed into hydroxyapatite by hydrothermal exchange. The porosity and pore size distribution of the attained material, which is highly dependent to the species of origin, provides an osteoconductive scaffold for bone growth, undergoing timely dissolution and resorption through the bone remodeling process. Evidence of successful periodontal regeneration has been attained with the use of these xenografts [34].

1.2.4. Alloplastic Materials

New-generation alloplastic materials are biocompatible, inorganic-derived, synthetic bone substitutes. They possess some of the desired mechanical qualities of bone as well as osteoconductive properties, but are largely reliant on neighboring viable periosteum/bone for the attainment of clinical success. They primarily function as defect fillers.

Alloplastic materials with relevance in periodontal applications can be classified as polymers and ceramics/glasses, the later including bioactive glasses, calcium phosphate cements, bioceramics and multiphasic materials.

1.2.5. Polymers

Polymers are potential candidates for periodontal bone grafting procedures and there is a wide range of available preparations with different physical, mechanical, and chemical properties. The polymers with clinical application can be loosely divided into natural polymers and synthetic polymers. These, in turn, can be sub-divided into degradable and nondegradable [35]. Natural polymers include polysaccharides (*e.g.* agarose, alginate, hyaluronic acid, chitosan) and polypeptides (*e.g.* collagen, gelatin). Regarding their successful clinical application, one may not overlook the limited structural properties of natural polymers (*i.e.* a comparatively weak mechanical strength and variable rates of degradation) that restrict their application as standalone grafting biomaterials. Synthetic polymers (*e.g.* poly (glycolic acid), poly (L-lactic acid), polyorthoester, polyanhydride), on the other hand, provide an easy manipulation of their properties, which can be effortlessly tailored into the desired characteristics of scaffolds for tissue engineering applications. In orthopedics, synthetic polymers have an extended use as injectable and solid products for bone tissue regeneration [36]. In periodontal applications, polymers have been more widely used as barrier materials, as part of guided tissue regeneration (GTR) applications, and their application for the regeneration of intrabony defects seems to be far more limited [37]. Recently, some clinical reports provide evidence for the effectiveness of a microporous polymer containing polymethylmethacrylate, polyhydroxylethylmethacrylate and calcium hydroxide, in the treatment of periodontal intraosseous defects [38, 39].

1.2.6. Bioactive Glasses

Bioactive glasses are silicone-based, osteoconductive materials that bind to bone through the formation of carbonated hydroxyapatite. When exposed to body fluids, bioactive glasses are expected to be covered by a double layer composed of silica gel and a calcium-phosphorous rich apatite layer. The latter promotes adsorption and concentration of proteins utilized by osteoblasts to form a mineralized extracellular matrix. It is believed that these bioactive properties guide and promote osteogenesis, allowing rapid formation of bone [40]. Periodontal clinical parameters assessment revealed an increase in the clinical attachment level and hard tissue fill, when intrabony defects were implanted with bioactive glass. Nonetheless, this biomaterial exhibits essentially osteoconductive properties and histologic analysis of human periodontal defects revealed that attained healing was based on connective tissue encapsulation of the graft material and epithelial down-growth – with minimal evidence of regenerated cementum or connective tissue attachment [41].

1.2.7. Calcium Phosphate Cements

In orthopaedic and bone tissue engineering applications, calcium phosphate cements are gaining special interest due to their biomimetic nature and potential use as controlled release systems. Common components of the powder besides tetracalcium phosphate, dicalcium phosphate dihydrate or anhydrous, include monocalcium phosphate monohydrate and anhydrous, octacalcium phosphate, tricalcium phosphate, hydroxyapatite and fluorapatite [42]. The liquid phase includes water, calcium- or phosphate-containing solutions, organic acids or aqueous solutions of polymers [42]. The primary role of the liquid is to provide a vehicle for the dissolution of the reactants and precipitation of the products. The fabrication of calcium phosphate cement is a versatile process which yields a variety of injectable pastes and set cement materials, with a wide range of physicochemical and mechanical properties [43]. These results are broadly dependent on the characteristics of the solid and the aqueous phase, as well as conditions in which the mixing reaction is conducted. One feature of particular interest in these cements is that they are intrinsically porous, with a relevant percentage of porosity within the nano to submicron range [43]. In periodontal applications, clinical implantation of calcium phosphate cements has reported controversial results, it has been shown to perform better than hydroxyapatite ceramic granules [44] and failed to demonstrate superior clinical outcome in comparision to open flap debridement [45]. A recent work showed that at 6 months, intrabony defects implanted with calcium phosphate cements revealed a probing depth reduction and a gain the clinical attachment level.

However, no site showed periodontal regeneration and there was no histological evidence of new bone formation. Moreover, the presence of new cementum and new organized connective tissue were residual [46].

1.2.8. Bioceramics

Bioceramic alloplasts have been the most used materials in periodontal regenerative approaches. They comprise mainly calcium phosphate materials, with a calcium/phosphorous relation similar to that of the human bone. The two most widely used forms are tricalcium phosphate and hydroxyapatite. These materials can be produced in either amorphous or crystalline phases, maintaining the same calcium and phosphorous ratios. Attained differences in the degree of crystalline arrangement are able to induce changes in physico-chemical and biologic characteristics of the developed ceramics.

Tricalcium phosphate includes both alpha- and beta- forms which are produced similarly, nonetheless presenting different resorption properties. The crystal structure of alpha tricalcium phosphate is monoclinic and consists of columns of cations while the beta tricalcium phosphate has a rhombohedral structure. Clinically, beta tricalcium phosphate is most commonly used, undergoing a relatively fast resorption within 6-18 months, following orthotopic implantion within the bone. Despite the promising results in orthopaedic applications [47], beta tricalcium phosphate has provided contradictory evidences in both animal and clinical trials of periodontal regeneration [48, 49]. Studies reported improvements in clinical measures (including reduction of the probing depth and gain in the clinical attachment level) after the treatment of intrabony periodontal defects with beta tricalcium phosphate, thus only minimal regeneration occurred, being the major portion of healing attained with the formation of a long junctional epithelium, with limited new connective tissue attachment [49]. Despite the ability of this material to allow for bone deposition and ingrowth in orthotopic sites, it has been shown to become broadly fibro encapsulated when placed in periodontal intrabony defects, failing to stimulate new bone growth, and to be retained (with residual graft particles present) following long implantation times [50].

Synthetic hydroxyapatite, with the chemical composition of $Ca_{10}(PO_4)_6(OH)_2$, has found widespread and successful clinical utilization in the enhancement of the periodontal bone ingrowth. Its degradation can be controlled by the amount of porosity and degree of sintering, thus originating both absorbable and nonabsorble forms of hydroxyapatite [51]. Dense non-resorbable hydroxyapatite grafts are osteophilic, osteoconductive and act primarily as an inert biocompatible bone defect fillers. Resorbable forms, with a characteristic slow resorption rate, act as a mineral reservoir of calcium and phosphorous and, at the same time, as a scaffold for timely bone ingrowth. Hydroxyapatite presents remarkable biocompatibility with little inflammatory response when implanted within connective and bone tissues. Histological evidence of new bone formation around and in the neighborhood of grafted porous hydroxyapatite fragments, with the presence of osteocytes, osteoblasts and a normal peripheral connective tissue without inflammatory reaction have been reported in the literature [52]. Recently, advances in hydroxyapatite presentations have led to the development of substituted hydroxyapatites, with *e.g.* magnesion, silicon, fluoride, that seem to report improved biomechanical and biological behavior in orthopaedic applications [53]. Nonetheless, little evidence of application of these materials to periodontal regeneration has been validated.

1.2.9. Multiphasic Materials

More recently, there has been a growing interest in the development of multiphasic calcium phosphate ceramics as scaffolding materials for bone regenerative applications. They seem to be more effective in bone repair/regeneration than pure phase materials (*e.g.* hydroxyapatite, tricalcium phosphate) and have a controllable degradation rate, thus favoring the timely bone regeneration and remodeling processes [54]. Clinical application of different multiphasic materials revealed the efficiency for bone filling, performance for bone reconstruction and efficacy for bone ingrowth [55-57]. Bonelike® is a novel marketed multiphasic material, with proven clinical success in the bone-related regenerative applications [58].

1.2.10. Bonelike®

Bonelike® is prepared by a liquid-phase sintering route, in which hydroxyapatite is reinforced with a glass of the P_2O_5-CaO system [58]. During the sintering process the glass reacts with hydroxyapatite, forming beta tricalcium phosphate, which is following partially transformed into alpha tricalcium phosphate, at higher temperatures. The relative proportions of the tricalcium phosphate phases in the final microstructure depend upon several experimental factors, including the glass content and composition [59]. Attained composites reveal an increased bioactivity and improved biomechanical behavior in comparison to hydroxyapatite, due to the reducing of the grain size and porosity during the liquid sintering process of the materials' preparation. The bioactivity of Bonelike® is determined by an optimal balance of the least soluble phase of hydroxyapatite and most soluble phase of tricalcium phosphates. Further, the incorporation of different ionic species, such as carbonate, magnesium, sodium and fluoride, result in the development of a material with a chemical composition most similar to the one of the mineral phase of the human bone. In terms of biomechanical properties, and comparing to hydroxyapatite, Bonelike® presents significantly higher values for flexural binding strength and fracture toughness [60].

Bonelike® grafts are currently being used in several successful clinical applications namely in orthopedic and oral/maxillofacial procedures. In orthopedic applications, it has been employed for the regeneration of several bone defects caused by trauma or ageing [61, 62]. Within the oral/maxillofacial area, Bonelike® has been used for the regeneration of maxillary and mandible bone, after cyst removal, impacted teeth extraction, for sinus lift and bone augmentation around implants, as well as for maxilla and mandible reconstruction [63-66]. Recently, it has been used with success in the regeneration of intrabony periodontal defects in a case-series of patients with aggressive periodontitis [67].

1.3. Clinical Management of Periodontal Defects

Successful bone graft therapy in periodontal application relies on the basic principles of periodontal surgical technique, including a thoughtful case/defect selection, cautious pre-operative and post-operative management, and a meticulously surgical technique. Some studies have investigated the possible sources of variability in the clinical success of bone grafting procedures in periodontal surgery and elected the following factors as determinant.

1.3.1. Patient Related Factors

Patient-related factors in the prognosis of the success of grafting periodontal regenerative interventions include the degree of plaque control, presence of residual periodontal infection, use of tobacco, drug administration, patient's compliance and presence of systemic-associated condition, including diabetes, hyperparathyroidism, thyrotoxicosis, osteomalacia, osteoporosis and Paget's disease [17, 68].

1.3.2. Morphology of the Defect

The characteristic morphology of the intrabony defect seems to greatly determine the prognosis of the regenerative approach. Horizontal patterns of alveolar bone loss seem to be marginal responsive to periodontal regenerative therapies, including bone graft implantation; while vertical or angular bony defects, counting furcation defects, are often responsive to periodontal regeneration. Among the factors related to the anatomy of the defect, depth of the intrabony component and probing depth are consistently found to be relevant, as well as the number of residual bony walls defining the defects with two and three bony walls respond more favorably to treatment than do one-wall defects [17, 68].

1.3.3. Selection of the Graft Material

Selection of grafting material should be guided by the following principles: biologic acceptability, clinical predictability, degree and rate of resorbability, clinical feasibility, minimal operative hazards, postoperative sequelae and patient acceptance [17, 33, 68]. Also, only materials with a particle size range between 125 μm and 1000 μm should be employed. Particles with less than 100 μm in size elicit a pro-inflammatory response and are readily taken by macrophages, being rapidly resorbed with little or no new bone formation [69].

1.3.4. Surgical Procedure

The surgical technique for the treatment of periodontal intrabony defects with bone replacement grafts is essentially the same regardless of the type of graft material being used. Incisions are designed to allow for primary closure of flaps to protect the graft site from infection and the graft material from displacement [70]. Broadly, intrasulcular incisions are the common choice, with emphasis on preserving interdental tissue. Full thickness flaps are reflected to expose the underlying osseous defects and allow access for a thorough debridement of the defects and meticulous root planing [70]. Complete defect debridement and root surface decontamination are essential before placement of the bone graft material.

Accordingly, the periodontal defect should be debrided of all soft tissue using hand, ultrasonic, and/or rotary instruments. Moreover, the meticulous removal of all hard and soft accretions on the root surface and any clinically affected cemental surface and root abnormalities is mandatory. Following root debridement and preparation of the defect site, the bone grafting material should be carefully implanted, lightly condensed, and contoured to mimic the normal architecture of the adjacent alveolar bone. After suturing, slight pressure on the facial and lingual flaps should be applied to minimize the clot beneath the flap. Following, a periodontal dressing should be placed to protect the wound, without displacement of the graft or compromising the blood supply to the gingival flaps.

Selection of a specific flap design, in relation to anatomical characteristics of interdental space and location/morphology of bony lesion, and proper suturing technique may significantly contribute in determining the amount of soft and hard tissue changes following surgery [71]. This was confirmed by a significant center-related effect on treatment outcome observed when a biomaterial/ regenerative therapeutic intervention has been evaluated in a multicenter trial [72].

2. CLINICAL CASE

2.1. Case Presentation

A male patient with aggressive periodontitis and the presence of two intrabony periodontal defects of similar morphology was enrolled into a split-mouth design, in order to address the regenerative capabilities of an alloplastic material (Bonelike®, from Medmat Innovation). In this design, the test site comprised the implantation of Bonelike® following surgical debridement, while the control site was treated only with surgical debridement.

Clinical and radiological data were recorded at baseline and following 6 months post-operatively, for both control and test site, and are presented on Table **1**. Pre-operative radiographic (Fig. **1**) and tomographic images (Figs. **2-4**) allowed for the computation of radiological parameters. Clinical parameters were assessed with periodontal probing (Fig. **5**).

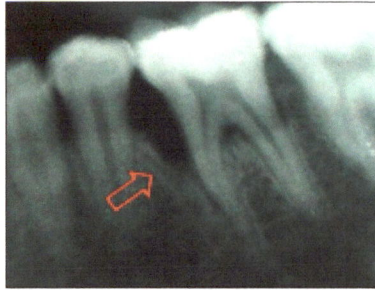

Figure 1: Preoperative radiographic image of the area around the mandibular left first molar, suggesting advanced bone loss at the mesial aspect (test site).

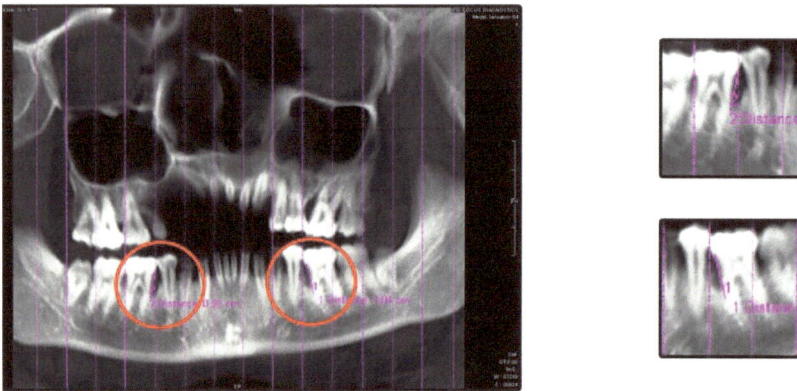

Figure 2: Preoperative Tomographic image showing the linear measurements on panoramic view - from cement-enamel junction (CEJ) to base of defect (BOD). The area of interest in the panoramic view is encircled and shown in the inset. Copyright Quintessence Publishing Co Inc.

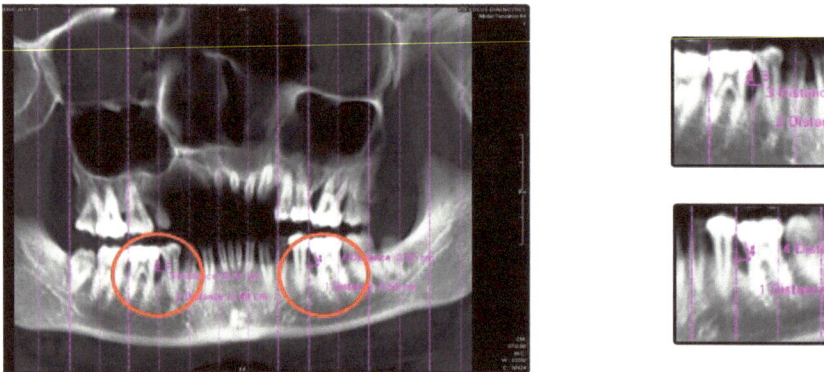

Figure 3: Preoperative Tomographic image showing the linear measurements on panoramic view- from CEJ to alveolar crest (AC). The area of interest in the panoramic view is encircled and shown in the inset. Copyright Quintessence Publishing Co Inc.

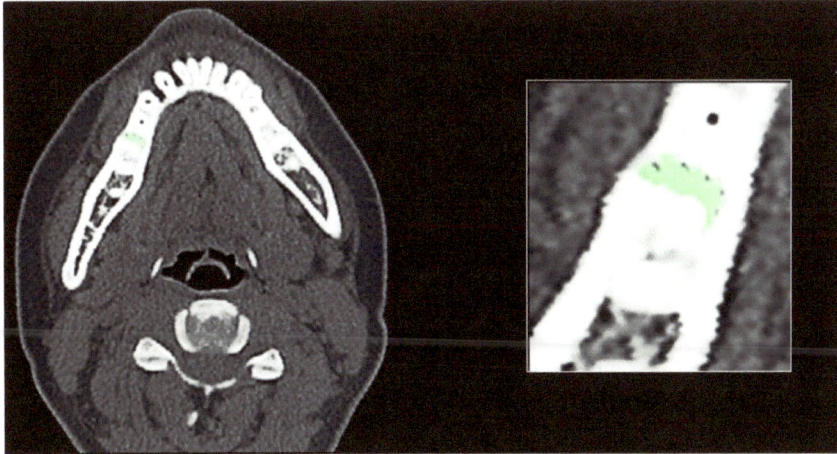

Figure 4: Mapping of the defects volume on axial section. The defect outline depicted in green color is shown in the inset. Copyright Quintessence Publishing Co Inc.

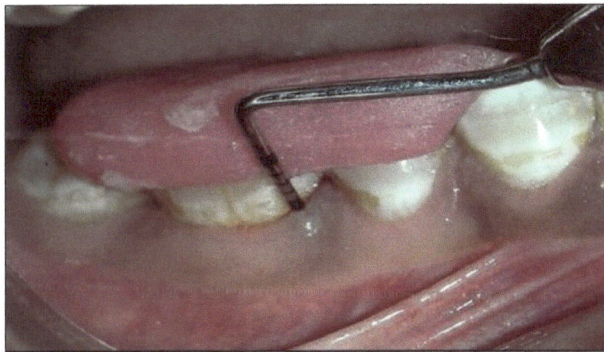

Figure 5: Preoperative measure of the probing depth using an acrylic stent and a UNC 15 periodontal probe.

2.2. Surgical Procedure

Routine periodontal pre-operative preparation was performed and local anesthesia was given. A crevicular incision was given which allowed for flap elevation, by means of blunt dissection with the help of a periosteal elevator. All the granulation tissue was carefully removed to ensure a clean site, and this was followed by thorough root planing. Further, during surgical protocol, utmost care was taken to preserve the interdental papilla (which aims to allow a better interproximal coverage of the graft material and to prevent exposure and exfoliation of the graft). An intra-operative image following defect debridement is shown in Fig. **6**.

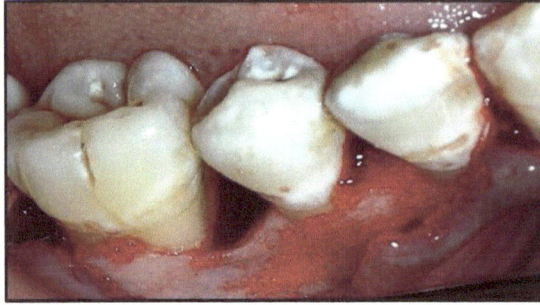

Figure 6: Operative view following the reflection of a full thickness flap (assessed by means of a sulcular incision) and defect debridement. The defect is visualized at the mesial aspect of the first molar.

Following, an adequate quantity of the graft material (Bonelike$^{®}$) was dispensed into a sterile dappen dish and mixed with an adequate quantity of patient's own blood drawn from the defect area. Before the graft implantation, a 3-0 non resorbable suture was passed through the buccal and lingual papillae and the suture was left loose (in order to prevent removal of the graft particles by the passage of the needle). At the test site, the alloplastic material was carried into the defect using a cumine scaler, in small increments, and filled to the level of the alveolar crest. The graft was then gently compacted and adapted to the anatomical contour of the alveolar bone. At the control site, the defect area was closed with sutures following surgical debridment alone. An intraoperative image of the test site is shown on Fig. **7**.

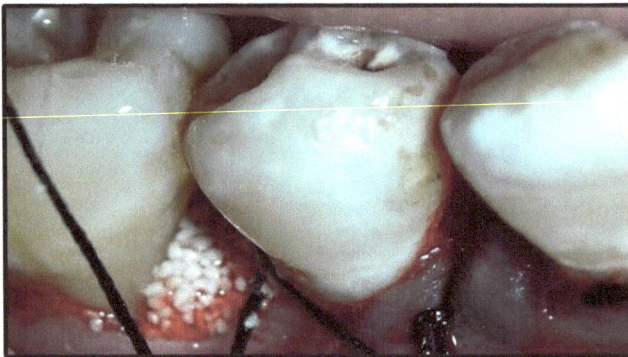

Figure 7: Intraoperative image of the allograft material contained within the defect.

The pre sutures placed were tightened and completed using routine interrupted interdental sutures, as shown in Fig. **8**. This allowed the repositioning and securement of the mucoperiosteal flaps.

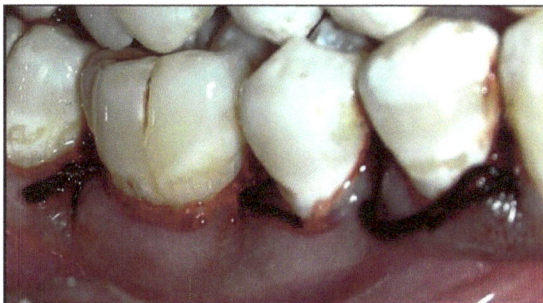

Figure 8: Intraoperative view of the sutured flap showing a complete coverage of the grafted location.

A non-eugenol based periodontal dressing (Coe Pak®, from GC Europe) was immediately placed at the surgical site, as shown in Fig. **9**.

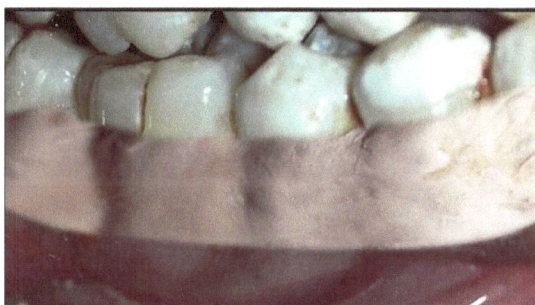

Figure 9: Intraoperative view of the placed periodontal dressing.

Routine postoperative instructions were carefully given to the patient and antibiotics and analgesics were prescribed, throughout the first 7 days following surgery. The patient was recalled 7 days after the surgical intervention and the periodontal dressings and sutures were removed. The patient was instructed to gently brush the area with a soft-bristled toothbrush. Patient was continuously monitored at 1, 3 and 6 months and, at each visit, oral hygiene was assessed and oral hygiene instructions were reinforced. At 6 months, all soft tissue measurements were repeated and a CT scan was taken to address hard tissue outcomes.

2.3. Results and Discussion

This clinical case aimed to address the response of periodontal intraosseous defects, treated by open flap debridement with and without the implantation of a glass reinforced hydroxyapatite alloplast (Bonelike®). The grafted material was well tolerated without any clinical signs of inflammation, infection or impaired healing.

Attained results revealed that both treatment modalities (surgical debridement, at the control site and bone graft implantation, at the test site) resulted in the reduction of the probing depth and in the gain of the clinical attachment level - the assessed clinical parameters associated with improved periodontal health. The reduction in the probing depth, from baseline till 6 months post-operatively was higher in the test group, comparing to control (4 mm *versus* 2 mm, respectively), as well as the gain in the clinical attachment level (4 mm *versus* 2 mm, respectively).

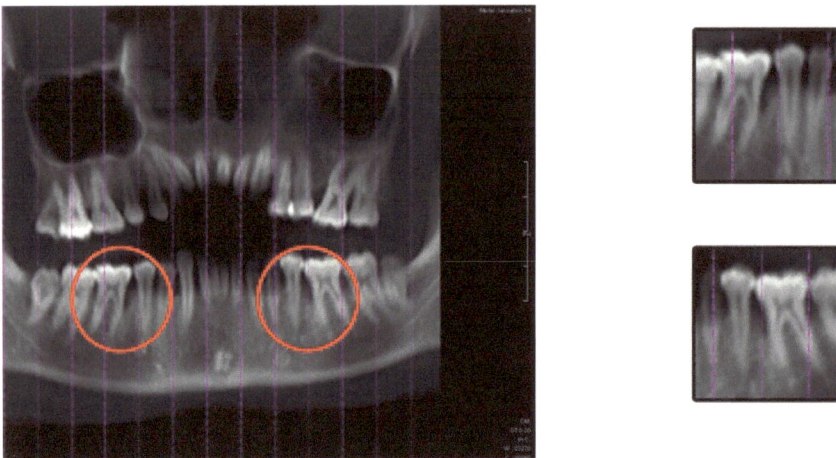

Figure 10: Post operative tomographic image showing the test and control sites six months after surgery. Areas of interest in the panoramic view are encircled and shown in the inset. Copyright Quintessence Publishing Co Inc.

Conducted tomographic analysis (Fig. **10**) also revealed a reduction in bone defect depth, from baseline to the final evaluated time point (6 month postoperatively), both at control and test sites (1.5 mm and 2.4 mm, respectively) and in the defect volume, which reduced 35.9 mm^3 in the control site and 56.3 mm^3 in the test site. Results are summarized in the Table **1**. Percentage of bone defect fill was calculated according to the following equation,

$$\frac{(Baseline\ bone\ volume\ -\ Final\ bone\ volume)}{Baseline\ bone\ volume} \times 100$$

and revealed a result of 58.56% for the control site and 68.57% for the test site.

Table 1: Clinical and radiological parameters at baseline (preoperatively) and 6 months postoperatively for the control and test sites

Parameter	Control site (baseline)	Control Site (6 months post-op.)	Test Site (baseline)	Test Site (6 months post-op.)
Probing depth (mm)	7	5	8	4
Clinical attachment level (mm)	8	6	9	5
Bone defect depth (mm)	4.5	3.0	5.2	2.8
Defect volume (mm^3)	61.3	25.4	82.1	25.8

Bone defect depth was calculated from the difference between CEJ – BOD measure and CEJ – AC measure, at the tomographic analysis, as shown in Figs. **2** and **3**.

These results demonstrated that the site treated with Bonelike$^®$ implantation showed an increased bone defect fill, comparing to the surgical debridement alone. Tomographic and clinical measures substantiate the attained findings. Bonelike$^®$ seems to adequately induce new bone tissue formation in intrabony periodontal defects, in a greater extent than surgical debridement alone. These results are in accordance with those published by Kumar *et al.* that showed the successful application of Bonelike$^®$ in the regeneration of periodontal bone defects [67, 73].

Several literature reports aimed to address the impact of periodontal bone graft procedures in the treatment/regeneration of periodontal intraosseous defects. Recently, two high evidence literature reports, meta-analysis assessed systematic reviews, have been published and aimed to address the proposed questions.

Trombelli *et al.* aimed to determine the adjunctive effect of grafting biomaterials with open flap debridement in the treatment of deep intraosseous defects [74]. In this systematic review, the authors addressed a wide range of studies implanting several bone grafting materials into periodontal intrabony defects. These included autologous bone grafts, bone allografts, coralline xenografts, bioactive glass, several formulations of hydroxyapatite of both synthetic and biological origin, in either porous and dense forms, polylactic acid and a polymethyl methacrylate/

polyhydroxyl-ethyl methacrylate composite. Apart from the implantation of polylactic acid, the implantation of all the other bone substitutes produced favorable changes in clinical attachment level and probing pocket depth – clinical measures of periodontal health - and increased defect fill when compared with open flap debridement alone [74]. Nonetheless, the authors verified a marked heterogeneity in the attained results from different studies, in each group of the assessed biomaterials.

The systematic review of Reynolds *et al.* aimed to address the efficacy of bone replacement grafts in proving demonstrable clinical improvements in periodontal osseous defects compared to surgical debridement alone [19]. The authors showed that bone graft implantation allowed for an increase in the bone level and in the clinical attachment level, and a reduction in the crestal bone loss and in the probing pocket depth, as compared to surgical debridement alone [19]. Further, they showed no significant differences in the clinical measurements between the use of synthetic hydroxyapatite and bone allografts and concluded that bone replacement grafts provide demonstrable clinical improvement of the periodontal bone regeneration, compared to surgical debridement alone [19].

CONCLUSIONS

There is substantial clinical and histological evidence that support the concept that autogenous bone grafts and demineralized freeze-dried bone allografts are effective regenerative materials in the treatment of periodontal intrabony defects. Moreover, several synthetic alloplastic materials have also been proven to be of clinical utility, enhancing the process of periodontal regeneration. In this work, Bonelike®, a glass-reinforced hydroxyapatite was implanted in an intrabony periodontal defect and has demonstrated an increased regeneration (assessed by clinical and radiological parameters), in comparison to surgical debridement alone. Despite the need for additional controlled clinical trials, preliminary data substantiates the applicability and prospective successful clinical application of Bonelike® in periodontal regenerative applications.

ACKNOWLEDGEMENT

Declared none.

CONFLICT OF INTEREST

The contents presented in the chapter have been carefully written based on the review from the references cited and the results obtained from the investigations carried out by the authors. Further, there is no conflict of interest with other people or organizations in respect of the present research work. Also, there is no financial support from any other organizations.

DISCLAIMER

Figs. **2**, **3**, **4** and **10** are published as a courtesy of Quintessence Publishing Co Inc. Fig are reproduced as in Kumar PG, *et al.* Quintessence Int 2011;5:42:375-384. Copyright Quintessence Publishing Co Inc.

REFERENCES

[1]   Pihlstrom B, Michalowicz B, Johnson N. Periodontal diseases. The Lancet 2005;366:1809-20.

[2]   Albandar J, Rams T. Global epidemiology of periodontal diseases: an overview. Periodontol 2000. 2002;29:7-10.

[3]   Borrell L, Burt B, Taylor G. Prevalence and trends in periodontitis in the USA: from the NHANES III to the NHANES, 1988 to 2000. J Dent Res 2005;84:924–30.

[4]   Albandar J. Epidemiology and risk factors of periodontal diseases. Dent Clin North Am 2005;49:517-32.

[5]   Genco R. Current view of risk factors for periodontal diseases. J Periodontol 1996;67:1041-9.

[6]   Albandar J. Global risk factors and risk indicators for periodontal diseases. Periodontol 2000. 2002;29:177-206.

[7]   Kinane D, Hart T. Genes and gene polymorphisms associated with periodontal disease. Crit Rev Oral Biol Med 2003;14:430-49.

[8]   Socransky S, Haffajee A, Cugini M, Smith C, Kent RJ. Microbial complexes in subgingival plaque. J Clin Periodontol 1998;25:134-44.

[9]   Moutsopoulos N, Madianos P. Low-grade inflammation in chronic infectious diseases: paradigm of periodontal infections. Ann N Y Acad Sci 2006;1088:251–64.

[10]   Socransky S, Haffajee A. Periodontal microbial ecology. Periodontol 2000. 2005;38:135-87.

[11]   Taubman M, Kawai T, Han X. The new concept of periodontal disease pathogenesis requires new and novel therapeutic strategies. J Clin Periodontol 2007;34:367-9.

[12]   Oringer R. Modulation of the host response in periodontal therapy. J Periodontol 2002;73:460-70.

[13]   Smith M, Seymour G, Cullinan M. Histopathological features of chronic and aggressive periodontitis. Periodontol 2000. 2010;53:45-54.

[14] Armitage G, Cullinan M. Comparison of the clinical features of chronic and aggressive periodontitis. Periodontol 2000. 2010;53:12-27.

[15] Conn C. Clinical significance of non-surgical periodontal therapy: an evidence-based perspective of scaling and root planing. J Clin Periodontol 2002;29:22-32.

[16] Kaldahl W, Kalkwarf K, Patil K, Molvar M, Dyer J. Long-term evaluation of periodontal therapy: I. Response to 4 therapeutic modalities. J Periodontol 1996;67:93-102.

[17] Caton J, Greenstein G. Factors related to periodontal regeneration. Periodontol 2000. 1993;1:9-15.

[18] Hanes P. Bone replacement grafts for the treatment of periodontal intrabony defects. Oral Maxillofac Surg Clin North Am 2007;19:499-512.

[19] Reynolds M, Aichelmann-Reidy M, Branch-Mays G, Gunsolley J. The efficacy of bone replacement grafts in the treatment of periodontal osseous defects. A systematic review. Ann Periodontol 2003;8:227–65.

[20] McAllister B, Haghighat K. Bone augmentation techniques. J Periodontol 2007;78:377–96.

[21] Nasr H, Aichelmann-Reidy M, Yukna R. Bone and bone substitutes. Periodontol 2000. 1999;19:74-86.

[22] Mellonig J. Autogenous and allogeneic bone grafts in periodontal therapy. Crit Rev Oral Biol Med 1992;3:333-52.

[23] Garrett S, Bogle G. Periodontal regeneration with bone grafts. Curr Opin Periodontol 1994;12:168-77.

[24] Kiyokawa K, Kiyokawa M, Hariya Y, Fujii T, Tai Y. Regenerative treatment of serious periodontosis with grafting of cancellous iliac bone and gingival flaps and replanting of patients' teeth. J Cranifac Surg 2002;13:375-81.

[25] Younger E, Chapman M. Morbidity at bone graft donor sites. J Orthop Trauma 1989;3:192-5.

[26] Gazdag A, Lane J, Glaser D, Forster R. Alternatives to Autogenous Bone Graft: Efficacy and Indications. J Am Acad Orthop Surg 1995;3:1-8.

[27] Mellonig J. Freeze-dried bone allografts in periodontal reconstructive surgery. Dent Clin North Am 1991;35:505-20.

[28] Finkemeier C. Bone-grafting and bone-graft substitutes. J Bone Joint Surg Am 2002;84A:454-64.

[29] Zhang M, Powers RJ, Wolfinbarger LJ. Effect(s) of the demineralization process on the osteoinductivity of demineralized bone matrix. J Periodontol 1997;68:1085-92.

[30] Schwartz Z, Mellonig J, Carnes DJ, de la Fontaine J, Cochran D, Dean D, *et al.* Ability of commercial demineralized freeze-dried bone allograft to induce new bone formation. J Periodontol 1996;67:918-26.

[31] Bender S, Rogalski J, Mills M, Arnold R, Cochran D, Mellonig J. Evaluation of Demineralized Bone Matrix Paste and Putty in Periodontal Intraosseous Defects. J. Periodontol 2005;76:768-77.

[32] AlGhamdi A, Shibly O, Ciancio S. Osseous grafting part II: xenografts and alloplasts for periodontal regeneration--a literature review. J Int Acad Periodontol 2010;12:39-44.

[33] Rosen P, Reynolds M, Bowers G. The treatment of intrabony defects with bone grafts. Periodontol 2000. 2000;22:88–103.

[34] Demers C, Hamdy C, Corsi K, Chellat F, Tabrizian M, Yahia L. Natural coral exoskeleton as a bone graft substitute: a review. Biomed Mater Eng 2002;12:15-35.

[35] Khan Y, Yaszemski M, Mikos A, Laurencin C. Tissue engineering of bone: material and matrix considerations. J Bone Joint Surg Am 2008;90 Suppl1:36-42.

[36]   Kretlow J, Mikos A. Review: mineralization of synthetic polymer scaffolds for bone tissue engineering. Tissue Eng 2007;13:927-38.

[37]   Sculean A, Nikolidakis D, Schwarz F. Regeneration of periodontal tissues: combinations of barrier membranes and grafting materials - biological foundation and preclinical evidence: a systematic review. J Clin Periodontol 2008;35(8 Suppl):106-16.

[38]   Yukna R. HTR polymer grafts in human periodontal osseous defects. I. 6-month clinical results. J Periodontol 1990;61:633-42.

[39]   Calongne K, Aichelmann-Reidy M, Yukna R, Mayer E. Clinical comparison of microporous biocompatible composite of PMMA, PHEMA and calcium hydroxide grafts and expanded polytetrafluoroethylene barrier membranes in human mandibular molar Class II furcations. A case series. J Periodontol 2001;72:1451-9.

[40]   Sohrabi K, Saraiya V, Laage T, Harris M, Blieden M, Karimbux N. An Evaluation of Bioactive Glass in the Treatment of Periodontal Defects: A Meta-Analysis of Randomized Controlled Clinical Trials. J Periodontol 2012;83(4):453-64.

[41]   Sculean A, Windisch P, Keglevich T, Gera I. Clinical and histologic evaluation of an enamel matrix protein derivative combined with a bioactive glass for the treatment of intrabony periodontal defects in humans. Int. J. Perio Rest Dent 2005;25:139-47.

[42]   Fukase Y, Eanes E, Takagi S, Chow L, Brown W. Setting reactions and compressive strengths of calcium phosphate cements. J Dent Res 1990;69:1852-6.

[43]   Ambard A, Mueninghoff L. Calcium phosphate cement: review of mechanical and biological properties. J Prosthodont 2006;15:321-8.

[44]   Rajesh J, Nandakumar K, Varma H, Komath M. Calcium phosphate cement as a "barrier-graft" for the treatment of human periodontal intraosseous defects. Indian J Dent Res 2009;20:471–9.

[45]   Shirakata Y, Setoguchi T, Machigashira M, Matsuyama T, Furuichi Y, Hasegawa K, *et al.* Comparison of injectable calcium phosphate bone cement grafting and open flap debridement in periodontal intrabony defects: a randomized clinical trial. J Periodontol 2008;79:25-32.

[46]   Mellonig J, Valderrama P, Cochran D. Clinical and histologic evaluation of calcium-phosphate bone cement in interproximal osseous defects in humans: a report in four patients. Int. J. Perio Rest Dent 2010;30:121-7.

[47]   Larsson S. Calcium phosphates: what is the evidence? J Orthop Trauma 2010;24 (Suppl 1):S41-S5.

[48]   Saffar J, Colombier M, Detienville R. Bone formation in tricalcium phosphate-filled periodontal intrabony lesions. Histological observations in humans. J Periodontol 1990;61:209–16.

[49]   Stavropoulos A, Windisch P, Szendröi-Kiss D, Peter R, Gera I, Sculean A. Clinical and histologic evaluation of granular beta-tricalcium phosphate for the treatment of human intrabony periodontal defects: a report on five cases. J Periodontol 2010;81:325-34.

[50]   Froum S, Stahl S. Human intraosseous healing responses to the placement of tricalcium phosphate ceramic implants. II. 13 to 18 months. J Periodontol 1987;58:103-9.

[51]   Klein C, Driessen A, de Groot K, van den Hooff A. Biodegradation behaviour of various calcium phosphate materials in bone tissue. J Biomed Mater Res 1983;17:769–84.

[52]   Benqué E, Gineste M, Heughebaert M. Histological study of the biocompatibility of hydroxyapatite crystals in periodontal surgery. J Biol Buccale 1985;13:271-82.

[53]   Matsumoto T, Okazaki M, Nakahira A, Sasaki J, Egusa H, Sohmura T. Modification of apatite materials for bone tissue engineering and drug delivery carriers. Curr Med Chem 2007;14:2726-33.

[54]   Daculsi G, Laboux O, Malard O, Weiss P. Current state of the art of biphasic calcium phosphate bioceramics. J Mater Sci Mater Med 2003;14:195-200.

[55]   LeGeros R, Lin S, Rohanizadeh R, Mijares D, LeGeros J. Biphasic calcium phosphate bioceramics: preparation, properties and applications. J Mater Sci Mater Med 2003;14:201-9.

[56]   Nair M, Suresh-Babu S, Varma H, John A. A triphasic ceramic-coated porous hydroxyapatite for tissue engineering application. Acta Biomater 2008;4:173-81.

[57]   Castellani C, Zanoni G, Tangl S, Van Griensven M, Redl H. Biphasic calcium phosphate ceramics in small bone defects: potential influence of carrier substances and bone marrow on bone regeneration. Clin Oral Implants Res 2009;20:1367-74.

[58]   Santos J, Hastings G, Knowles J. Sintered hydroxyapatite compositions and method for the preparation thereof. European Patent 1999(WO 0068164).

[59]   Prado-da-Silva M, Lemos A, Gibson I, Ferreira J, Santos J. Porous glass reinforced hydroxyapaite materials produced with different organic additives. J Non-Cryst Solids 2002;304:286-92.

[60]   Lopes M, Knowles J, Santos J. Structural insights of glass-reinforced hydroxyapatite composites by Rietveld refinement. Biomaterials 2000;21:1905-10.

[61]   Gutierres M, Dias A, Lopes M, Hussain N, Cabral A, Almeida L, *et al.* Opening wedge high tibial osteotomy using 3D biomodelling Bonelike macroporous structures: case report. J Mater Sci Mater Med 2007;18:2377-82.

[62]   Gutierres M, Lopes M, Sooraj Hussain N, Lemos A, Ferreira J, Afonso A, *et al.* Bone ingrowth in macroporous Bonelike® for orthopaedic applications. Acta Biomaterialia 2008;4:370-7.

[63]   Lobato J, Sooraj Hussain N, Botelho C, Maurício A, Lobato J, Lopes M, *et al.* Titanium dental implants coated with Bonelike®: Clinical case report. Thin Solid Films 2006;515:279-84.

[64]   Duarte F, Santos J, Afonso A. Medical applications of Bonelike in Maxillofacial Surgery. Mater Sci Forum 2004;370:455-6.

[65]   Oliveira M, Sooraj Hussain N, Dias A, Lopes M, Azevedo L, Zenha H, *et al.* 3-D biomodelling technology for maxillofacial reconstruction. Mater Sci Eng C 2008;28(8):1347-51.

[66]   Sousa R, Lobato J, Maurício A, Hussain N, Botelho C, Lopes M, *et al.* A Clinical Report of Bone Regeneration in Maxillofacial Surgery using Bonelike® Synthetic Bone Graft. J Biomat Appl 2008 January 1, 2008;22:373-85.

[67]   Kumar P, Kumar J, Anumala N, Reddy K, Avula H, Hussain S. Volumetric analysis of intrabony defects in aggressive periodontitis patients following use of a novel composite alloplast: a pilot study. Quintessence Int 2011;42:375-84.

[68]   Aichelmann-Reidy M, Reynolds M. Predictability of clinical outcomes following regenerative therapy in intrabony defects. J Periodontol 2008;79:387-93.

[69]   AlGhamdi A, Shibly O, Ciancio S. Osseous grafting part I: autografts and allografts for periodontal regeneration–a literature review. J Int Acad Periodontol 2010;12:34–8.

[70]   Cortellini P, Bowers G. Periodontal regeneration of intrabony defects: an evidence-based treatment approach. Int. J. Perio Rest Den 1995;15:128-45.

[71]   Trombelli L, Bottega S, Zucchelli G. Supracrestal soft tissue preservation with enamel matrix proteins in treatment of deep intrabony defects. J Clin Periodontol 2002;29:433-9.

[72]   Tonetti M, Lang N, Cortellini P, Suvan J, Adriaens P, Dubravec D, *et al.* Enamel Matrix Proteins in the regenerative therapy of deep intrabony defects. A multicentre randomized controlled clinical trial. J Clin Periodontol 2002;29:317–25.

[73]   Kumar P, Kumar J, Reddy K, Hussain N, Lopes M, Santos J. Application of Glass Reinforced Hydroxyapatite Composite in the Treatment of Human Intrabony Periodontal Angular Defects – Two Case Reports. Solid State Phenomena 2010;161:93-101.

[74]   Trombelli L, Heitz-Mayfield L, Needleman I, Moles D, Scabbia A. A systematic review of graft materials and biological agents for periodontal intraosseous defects. J Clin Periodontol 2002;29 Suppl 3:117-35.

S   *Send Orders of Reprints at bspsaif@emirates.net.ae*

CHAPTER 6

Physical, Optical and Structural Properties of Er³⁺ Doped Zinc/Cadmium Bismuth Borate/Silicate Glasses

Inder Pal[1], Ashish Agarwal[1,*], Sujata Sanghi[1] and Mahender P. Aggarwal[2]

[1]Department of Applied Physics, Guru Jambheshwar University of Science & Technology, Hisar-125001 Haryana, India and [2]Guru Nanak Khalsa College, Yamunanagar-135001, Haryana, India

Abstract: Glasses with compositions $20MO \cdot xBi_2O_3 \cdot (79.5-x)B_2O_3$ ($15 \leq x \leq 35$, x in mol%,) and $20MO \cdot xSiO_2 \cdot (79.5-x)Bi_2O_3$ ($10 \leq x \leq 50$, x in mol%, M = Zn and Cd) containing 0.5 mol% Er^{3+} ions were prepared by normal melt-quench technique ($1150°C$ in air). The effect of host glass composition on the optical absorption and fluorescence spectra of Er^{3+} ions have been observed with varying contents of Bi_2O_3. Judd-Ofelt approach has been applied for *f-f* transition of Er^{3+} ions to evaluate various intensity parameters *viz.* Ω_λ (λ = 2, 4, 6). The variation of Ω_2 with Bi_2O_3 content has been attributed to change in the asymmetry of ligand field at the rare earth ion site. The Judd-Ofelt intensity parameters were determined from intensities of absorption bands in order to calculate the radiative properties *viz.* transition probability (A_{rad}), and radiative life-time of the excited states (τ_r). From the emission spectra, full width at half maxima (FWHM), stimulated emission cross-section (σ) and figure of merit were evaluated and compared with other hosts. The observed NIR emission ($^4I_{13/2} \rightarrow {}^4I_{15/2}$ at 1.5μm) in Er^{3+}-doped zinc/cadmium bismuth borate/ silicate glasses may be useful in optical communication.

Keywords: Rare-earth ions, Rare-earth doped glasses, Laser Glasses, Oxide glasses, Heavy-metal oxide glasses, Judd-Ofelt theory, Optical material, Optical transition, Absorption spectra, Fluorescence spectra, Physical properties of glasses, FTIR, Radiative transition probability, Radiative lifetime, Stimulated emission cross-section, Figure of merit, Intensity parameters, Melt-quech technique, Matrix elements, Er³⁺ doped glasses.

1. INTRODUCTION

Rare earth ions have a long history in optical and magnetic applications. During the last four decades, there has been a rapid development of rare earth containing materials for a large number of optical applications such as lasers, phosphors,

---
*Address correspondence to Ashish Agarwal: Department of Applied Physics, Guru Jambheshwar University of Science & Technology, Hisar-125001, Haryana, India. Tel: +91-1662-263384, Fax: +91-1662-276240. Email: aagju@yahoo.com

Sooraj H. Nandyala and José D. Santos (Eds)
All rights reserved-© 2013 Bentham Science Publishers

infrared to visible upconvertors, quantum counters, communication systems, luminescent materials, solar concentrators, optical detectors, waveguide laser, optical fiber, wavelength division multiplexing (WDM) devices, *etc*. Rare earths have important characteristics that distinguish them from other optically active ions. For any optical application, the optimization of physical, chemical, thermal, optical and fluorescence properties of the rare earth doped materials is required. In order to meet the diversified requirements of optical materials, not only new materials have been developed but several existing materials have also been modified. Today rare earth doped glasses are not limited to infrared optical devices but there is growing interest in visible optical systems. Also, the rare earth doped glasses are attractive host because planar waveguides and optical fibers can be fabricated easily in comparison to crystalline materials. Oxide glasses are stable host for obtaining efficient luminescence in rare earth ions.

The term "rare earths" was suggested by Johann Gadolin in 1794: "rare" because when the first of the rare earth elements was discovered they were thought to be present in the earth's crust only in small amounts, and "earths" because as oxides they have an earthy appearance. The rare earths are divided into two groups of 14 elements each (Lanthanides and Actinides) and they form a group of chemically similar clements. Rare earths possess the common feature of accommodating 54 electrons in the xenon structure ($1s^2\ 2s^2\ 2p^6\ 3s^2\ 3p^6\ 3d^{10}\ 4s^2\ 4p^6\ 4d^{10}\ 5s^2\ 5p^6$) and the two outer electrons in $6s^2$ shell (except for the rare earth Ce, Gd and Lu in which there are three outer electrons in $5d6s^2$ shells); remaining electrons occupy the inner $4f$ shell [1] that distinguish them from other optically active ions such as transition metals. In condensed matter, the trivalent (3+) level of ionization is the most stable state for lanthanide ions, and most of optical devices use trivalent ions. In general, the trivalent rare earth ions are denoted by RE^{3+}. From application view point, the trivalent state of the rare earth ions is most important [2]. The electronic configuration of RE^{3+} ions have the basic xenon like structure of 54 electrons followed by N number of electrons in the $4f$ shell, with the value of N ranging from $1(Ce^{3+})$ to $14\ (Lu^{3+})$. The $4f$ electrons are shielded by $5s^2$ and $5p^6$ closed shells.

The electronic transitions within the $4f$ shell give sharp line spectra; some of the fluorescence transitions are identified as laser transitions. Professor Dieke's research group at Johns Hopkins (1963's) compiled a table of energy levels for the trivalent rare earths in crystals. The energy levels of free trivalent RE ions up to 42000 cm^{-1} measured by Dieke *et al.* [3] are shown in Fig. **1** (Dieke diagram). This diagram plays a very important role in analyzing the 4*f*–4*f* absorption spectra of trivalent RE ions in a free state and in crystals, and it was quite sufficient until recently.

Er^{3+} is one of the first paramagnetic ions studied for optical coherent transients with the possible exception of ruby. The 1.5 μm transition falls within the important telecommunications band and has thus attracted a lot of attention. The same transition is utilized in erbium doped amplifiers in telecommunications. The $^4I_{15/2}$ to $^4S_{3/2}$ transition at 18,557 cm^{-1} in Er^{3+}: LaF_3 was studied by Macfarlane and Shelby [4] in 1982 and the $^4I_{15/2}$ to $^4F_{9/2}$ for Er^{3+}: $YAIO_3$, Er^{3+}: $LiYF_4$, and Er^{3+}: LaF_3 around 649 nm was studied by Ganem *et al.* [5] in 1991. The emission at 1.5 μm in Er^{3+}: $^4I_{15/2} \rightarrow {}^4I_{13/2}$ transition was first studied by Silberberg *et al.* [6] in 1992 in Er^{3+} doped fibers. The Er^{3+} doped glasses and fibers have the potential to be applied in wide band internet because of inhomogeneous broadening for erbium ions in glass matrix [7]. Therefore, a great attention has been devoted to the research on Er^{3+} ions doped glasses [8-10]. Bismuth based glasses have high refractive index and have wide broadening effect on emissions of Er^{3+}. Recent investigation has shown that approximately 80 nm gain band at 1.55 μm was obtained in the Er^{3+} doped bismuth based glasses [11, 12], which have good chemical-physical properties that are favorable for the mechanical processing and for drawing fiber. Hence, the bismuth based oxides glasses containing Er^{3+} ions are suitable candidates for optical fiber amplification and optical fiber laser. Among the numerous host glasses, the oxide glasses doped with Er^{3+} ions are preferred as they possess low melting temperature, large refractive index, good physical and chemical properties *etc.* Saisudha *et al.* [13, 14] have investigated the effect of glass matrices on the spectroscopic properties of Nd^{3+}, Sm^{3+} and Dy^{3+} ions and reported that bismuth borate glasses may also act as one of the potential laser host materials. Theoretical analysis of spectroscopic properties of Er^{3+} ions in many glasses have been widely studied using the Judd-Ofelt theory [15, 16]. By changing the host glass compositions, the spectroscopic parameters of Er^{3+} ions that are used to estimate the laser performances of the glass can be modified.

Several authors have recently studied the properties of undoped [17], Er^{3+} doped [18], Nd^{3+} doped [19] bismuth borate glasses, however, in a restricted composition range. Becker could show that the refractive index can be varied [17]; Chen *et al.* proved that high luminescence efficiencies are possible especially in Nd^{3+} doped glass [20]. Tanabe *et al.* [21] showed that in Er^{3+} doped $Bi_2O_3 \cdot B_2O_3 \cdot SiO_2$ glasses, the luminescence lifetime and efficiency decrease with increasing B_2O_3 content. Similar results were also reported by Yang *et al.* [18] for $Bi_2O_3 \cdot B_2O_3 \cdot Na_2O$ glasses. Authors have recently investigated the optical properties of Sm^{3+} and Pr^{3+} ions in $ZnO \cdot Bi_2O_3 \cdot B_2O_3$ (ZBS and ZBP) glasses [22, 23]. Recently, Paul *et al.* [24] also studied the performance of zirconium and conventional bismuth based Er-doped fiber amplifiers (*EDFA*). They show that the flat-gain values, bandwidth and noise level are better in Zr-EDFA in comparison with Bi-EDFA.

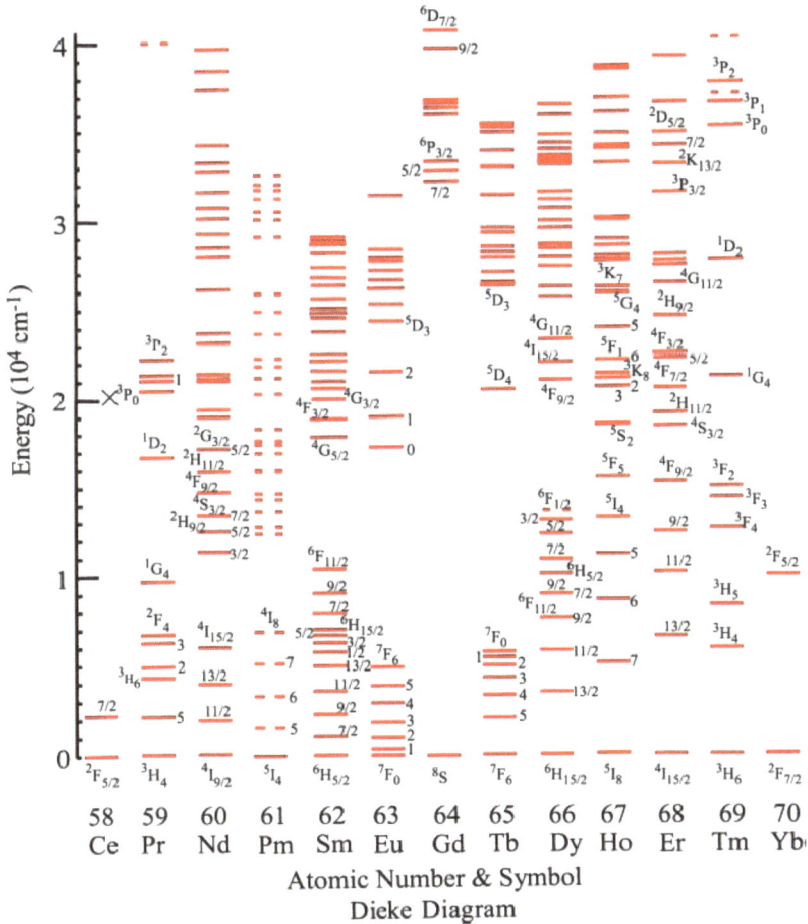

Figure 1: Dieke diagram (energy levels of free RE^{3+} ions up to 42000 cm^{-1}).

In view of above, in the present chapter Er^{3+} ions doped zinc/cadmium bismuth borate/silicate glass systems have been prepared using normal melt-quench technique and a systematic study has been carried out to observe the effect of unconventional glass former (Bi_2O_3) on physical, optical and fluorescence properties of these glasses. Further, an attempt has been made to correlate these properties with the structural changes in the samples.

2. EXPERIMETAL DETAILS

Glass samples have been prepared by using the standard melt-quench technique. In this technique glass is formed by the continues hardening (*i.e.* increase in viscosity) of the melt. For this, cooling rate was kept fast to preclude crystal nucleation and growth. Analar grade reagents were used as starting materials. The required amounts of chemicals were weighed in a single pan balance having accuracy of 10^{-2} mg. The weighed chemicals were properly mixed and then transferred to a porcelain crucible. For melting the mixture, crucible was placed in rapid heating electrical muffle furnace. The electrical muffle furnace has two units. The first unit is called heating unit which consists of a cavity of size 15.0 x 15.0 x 30.0 cm^3. The silicon carbide rods are used as heating elements at the top of cavity. On passing electric current through these rods, the temperature of the furnace can be raised up to 1773K. The second unit of the furnace is called the controlling unit. It controls the temperature of the furnace and consists of a strong power supply, which can provide electric current up to 40A. To melt the mixture of chemicals, the temperature of the furnace is set at 1473K and the melting time is about half an hour. To ensure homogeneity of the glass, the melt was regularly stirred during melting. Coin shaped glass samples of $\approx$ 0.5 to 1.5 mm thickness were obtained by pouring the melt onto a stainless steel plate and pressing with another. The prepared samples were then well polished to optical quality for spectral and other investigations. Prior to use, the samples thus formed were kept in a desiccator to protect them from atmospheric moisture. The exact compositions of the prepared glasses are given below and a sample code for each glass is presented in Table **1**.

(i) $20ZnO \cdot xBi_2O_3 \cdot (79.5-x)B_2O_3 \cdot 0.5Er_2O_3$ ($15 \leq x \leq 35$ mol%)   (ZBBE glasses)

(ii) $20CdO \cdot xBi_2O_3 \cdot (79.5-x)B_2O_3 \cdot 0.5Er_2O_3$ ($15 \leq x \leq 35$ mol%)   (CBBE glasses)

(iii) $20ZnO \cdot xSiO_2 \cdot (79.5-x)Bi_2O_3 \cdot 0.5Er_2O_3$ ($10 \leq x \leq 50$ mol%)   (ZSBE glasses)

(iv) $20CdO \cdot xSiO_2 \cdot (79.5-x)Bi_2O_3 \cdot 0.5Er_2O_3$ ($10 \leq x \leq 50$ mol%)   (CSBE glasses)

We obtained clear, bubble free and transparent glasses for all the compositions. Disk shape samples were ground and polished with different grades of emery powder. Glasses were characterized by density measurement, X-ray diffraction, UV-visible-NIR absorption spectroscopy, fluorescence spectra and Fourier Transform Infrared (FTIR) spectroscopy. The density (D) of each glass sample was measured by the Archimede's principle using Xylene as immersing liquid. The density was calculated according to the formula:

$$D = \frac{W_{air}}{W_{air} - W_{xylene}} \times \rho_{xylene} \qquad (1)$$

where W_{air} and W_{xylene} are the weight of the glass sample in air and in xylene, respectively and ρ_{xylene} is the density of xylene ($= 0.8645$ gcm^{-3}). All measurements were repeated two to three times per sample. The accuracy of results in density measurement is ± 0.001 gcm^{-3}. The molar volume was calculated from the traditional relation $V_M = M/D$ where M is the molar mass of the glass. The refractive index (n) measurements were carried out on polished glass samples. Brewster angle technique was used for measurement of n using He-Ne laser (632nm) as a source. The amorphous nature of the prepared samples was confirmed by recording X-ray diffraction using Cu Kα X-ray radiation, λ=1.5406 Å, in the 2θ range of 10-70^0 on Rigaku equipment using rotating anode. The X-ray tube was operated at 40kV and 30mA.

The optical absorption spectra of all the well polished glass (1.10-1.78 mm) samples were recorded at room temperature in the wavelength range from 300 to 3200 nm using a Varian (Carry 5000) spectrophotometer. The absorption coefficient as a function of wavelength, $\alpha(\lambda)$, was calculated by dividing the measured absorption by sample thickness. The emission spectra (in visible as well as near-infrared region) of polished samples were recorded on Jobin Yvon Flurolog-3-11 spectrofluorometer with a resolution limit of 0.2 nm using Xe-arc lamp (450W) as the excitation source at a wavelength of 380 nm and 980 nm, respectively. IR transmission spectra of the glasses were recorded at room temperature (RT) using KBr pellet technique on a Shimadzu FTIR 8001PC spectrometer in the range 400–4000 cm^{-1}. The samples investigated were fine particles mixed with pulverized KBr in the ratio 1:20 mg glass powder to KBr, respectively. The weighed mixture was then subjected to a pressure of 7–8 tons to produce clear homogenous discs.

3. OSCILLATOR STRENGTH

3.1 Experimental Determination

The experimental value of oscillator strength, f_{expt}, of an electronic transition can be deduced utilizing the optical absorption data. It is given by the expression [25]:

$$f_{expt} = 4.32 \times 10^{-9} \int \varepsilon(\bar{v}) d\bar{v} \qquad (2)$$

where $\varepsilon(\bar{v})$ is the molar absorptivity (cm^2) of absorption band at energy $\bar{v}$ (cm^{-1}). The molar absorptivity at a given energy can be computed from the well known Beer-Lambert's law, *i.e.*

$$\varepsilon(\bar{v}) = \frac{1}{Cl} \log(I_0/I) \qquad (3)$$

where C is the concentration (moles/1000cm^3) of the rare earth ions, l is the thickness (cm) of the sample through which light passes and $\log(I_0/I)$ is the absorbance (or optical density). The value of absorbance can be obtained from the experimentally observed spectrum. It is a good approximation [26] to write a Gaussian error-cure for $\varepsilon(\bar{v})$, *i.e.*

$$\varepsilon(\bar{v}) = \varepsilon_{max} e^{-\left(\frac{\bar{v}-\bar{v}_0}{W}\right)^2 4\ln 2} \qquad (4)$$

where $\bar{v}_0$ is the centroid of the absorption peak, W is the full width at half maximum and ε_{max} is the molar absorptivity at the centriod of the peak. The value of ε_{max} can be calculated using the expression:

$$\varepsilon_{max} = \frac{1}{Cl}\left[\log\left(I_0/I\right)\right]_{max} \quad (\text{at } \bar{v} = \bar{v}_0)$$ (5)

From Eqs. (2) and (4), we have at $\bar{v} = \bar{v}_0$

$$f_{expt} = 4.32\times10^{-9}\int\varepsilon_{max}\,e^{-\left(\frac{\bar{v}-\bar{v}_0}{W}\right)^2 4\ln2}\,d\bar{v} = 4.32\times10^{-9}\times1.0645\,\varepsilon_{max}\,W$$

$$= 4.60\times10^{-9}\,\varepsilon_{max}\,W$$ (6)

Thus by substituting the values of ε_{max} and W for different absorption bands in Eq. (6), the f_{expt} value for the corresponding electronic transition can be estimated.

3.2 Theoretically Calculated Oscillator Strength: Judd-Ofelt Theory

Optical transitions in rare earth ions have been found to be predominantly electric in origin [27]. For electric dipole transitions the well known selection rules [28] are $\Delta l = \pm1$, $\Delta S = 0$ and $|\Delta L|$, $|\Delta J| \leq 2$ (except $0 \rightarrow 0$). Since $l = 3$ for the 4f-shell, all the levels of this shell are with odd parity. But according to Laporte's rule: a change of parity between the two levels is necessary for the electric dipole transition to be effective; the electronic transitions within the 4f-shell are therefore parity forbidden. The electric dipole transitions can however be allowed if the admixing of states of opposite parity is produced by the local field. It is possible if the rare earth ion is located at the non-centrosymmetric site (a site with no inversion symmetry) of the host crystallites. If a small admixture of $4f^{N-1}5d$ opposite parity state is present in the $4f^N$ configuration, then electric dipole transitions become allowed. These crystal field induced transitions are therefore termed as forced electric dipole transitions. Although magnetic dipole and electric quadrupole transitions are allowed by the selection rules, their contributions to the intensity of transitions are negligibly small [27]. These are not included in the present investigations.

Judd [15] and Ofelt [16] independently derived expressions for the oscillator strength of the crystal field induced electric dipole transitions of rare earth ions. Since their results were similar and also published simultaneously, the basic theory is known as Judd-Ofelt theory. The Judd-Ofelt model has been responsible for the increased understanding of the spectra of rare earth ions in a variety of hosts. It leads to an easy and rapid determination of the optical properties of rare earth ions. The first report on application of Judd-Ofelt theory to the glassy hosts was published by Krupke [29]. The disagreement between the results obtained from experiments and those calculated by applying Judd-Ofelt theory was less than 10 percent.

According to the Judd-Ofelt theory the oscillator strength for an electronic transition between an initial manifold [(S,L)J] and a final manifold [(S',L')J'] is given by the expression [30]:

$$f_{cal} = \frac{8\pi^2 mv}{3h(2J+1)}\left[\frac{(n^2+2)}{9n}S_{ed}+nS_{md}\right] \tag{7}$$

where n is the refractive index of the host matrix, v is the frequency of the transition, m is the mass of electron, h is the Planck's constant, J is the total angular momentum quantum number of the initial level and

$$S_{ed}\left[(S,L)J,(S',L')J'\right]= \sum_{\lambda=2,4,6} \Omega_\lambda \left|\left\langle(S,L)J\|U^\lambda\|(S',L')J'\right\rangle\right|^2 \text{ and}$$

$$S_{md}\left[(S,L)J,(S',L')J'\right]= \sum_{\lambda=2,4,6} \Omega_\lambda \left|\left\langle(S,L)J\|L+2S\|(S',L')J'\right\rangle\right|^2.$$

S_{ed} and S_{md} represent the line strengths for the induced electric dipole transitions and magnetic dipole transitions, respectively. Ω_λ ($\lambda = 2, 4, 6$) are the Judd-Ofelt intensity parameters and the terms $\|U^{(\lambda)}\|^2$ are the doubly reduced matrix elements of a unit tensor operator of rank λ calculated in the intermediate coupling approximation [15]. The matrix elements of unit tensor operator have been numerically evaluated by Carnall *et al.* [31, 32] for different transitions of rare earth ions in the aqueous solutions.

The Judd-Ofelt intensity parameters Ω_λ ($\lambda = 2, 4, 6$) are defined by the expression [30]:

$$\Omega_\lambda =(2\lambda+1) \sum_{t=1,3,5,7} (2t+1)^{-1}\left|A_{tp}\right|^2\Xi^2(t,\lambda) \tag{8}$$

Here A_{tp} (t, odd) are the odd parity terms in the static crystal field expansion and depend upon the site symmetry of the rare earth ion in a given host matrix. The quantity Ξ (t, λ) contains integrals involving the radial parts of the $4f^N$ wave functions and the excited opposite parity electronic state wave functions, and the energies separating these states. Since these terms are strongly host dependent, the intensity parameters Ω_λ are very sensitive to the host matrix. It is important to mention that a parameter Ω_λ with the same value of λ as of the highest $\|U^{(\lambda)}\|^2$ is of dominance. For example, if a particular transition is characterized by a large $\|U^{(\lambda)}\|^2$, the parameter Ω_λ will be dominant.

The values of Judd-Ofelt intensity parameters Ω_2, Ω_4, and Ω_6 can be obtained by a least square fitting method. In this method, the square of the differences between the experimentally measured values of oscillator strength, f_{expt}, and those calculated on the basis of the theoretical considerations, f_{cal}, are minimized to obtain the best set of intensity parameters. The quantity of agreement between the theoretically calculated and experimentally measured values of oscillator strengths can be expressed by the smallness of root mean square (r.m.s.) deviation between them. It is defined by the expression [30]:

$$\partial_{r.m.s.} = \sqrt{\frac{\Sigma\left(f_{cal} - f_{expt}\right)^2}{N-3}} \tag{9}$$

where N is the number of absorption bands utilized in the determination of three Ω-parameters and summation is taken over all the N bands. The attractive features of the Judd-Ofelt approach is that once the set of three intensity parameters are determined for a given host matrix, the probability of transition between any two levels of $4f^N$ configuration of the rare earth ions in that host matrix can be calculated. The optical properties of the rare earth ion, with the help of these intensity parameters, in a given host matrix can be deduced as explained in the next section.

4. OPTICAL PROPERTIES

The optical properties of active rare earth lasing ion are used to predict the performance of a laser system, for they determine the energy storage, extraction and overall efficiency of the system. Since the optical properties of rare earth ions are sensitive to the composition, it is of significance to study these properties for different host matrices to tailor the optimum laser performance. The important optical properties are discussed briefly in the following subsections.

4.1. Spontaneous Emission Probability

The spontaneous emission probability or the transition probability for the radiative decay, A_{rad} (s^{-1}), is the most significant optical parameters, for it controls the intensity of any fluorescent transition. It is also used in the determination of other optical properties of the rare earth ions. The spontaneous emission probability for a laser transition from an initial manifold [(S,L)J] to a final manifold [(S',L')J'] can be computed using the expression [29]:

$$A_{rad}\left[(S,L)J;(S',L')J'\right] = \frac{64\,\pi^4 e^2 \bar{v}^3}{3h(2J+1)} \frac{n\left(n^2+2\right)^2}{9}$$

$$\times \sum_{\lambda=2,4,6}\left|\left\langle(S,L)J\left\|U^{(\lambda)}\right\|(S',L')J'\right\rangle\right|^2$$

(10)

where J is the total angular momentum quantum number of the initial manifold and $\|U^{(\lambda)}\|^2$ are the matrix elements of unit tensor operator for the fluorescence transitions from metastable laser level to the various lower lying laser levels.

4.2. Radiative Lifetime

Lifetime of the metastable laser level is an important optical property, for it determines the energy storage efficiency of a laser system. For good efficiency, the lifetime should be longer than that of the pump pulse. The radiative lifetime of the metastable level is given by the expression [29]:

$$\tau_r = \frac{1}{\sum_{S',L',J'} A\left[(S,L)J;(S',L')J'\right]}$$

(11)

where the summation is over all the lower level manifolds.

4.3. Fluorescence Branching Ratio

Wherever more than one emission transitions occur from a given metastable level, the relative intensity of each transition can be estimated from the fluorescence branching ratio, β_r. A larger value of β_r corresponds to the higher relative intensity of an emission transition. β_r for a transition from metastable [(S,L)J] to lower manifold [(S',L')J'] is given by [29]:

$$\beta_r\left[(S,L)J;(S',L')J'\right] = \frac{A_{rad}\left[(S,L)J;(S',L')J'\right]}{\sum_{S',L',J'} A_{rad}\left[(S,L)J;(S',L')J'\right]} \tag{12}$$

where the sum is over all possible lower manifolds.

4.4. Stimulated Emission Cross-Section

The stimulated emission cross-section plays an important role in determining the energy extraction efficiency of a lasing transition. The transitions with large stimulated emission cross-section are, in general, characterized as good laser transitions. The peak stimulated emission cross-section, σ (cm^2), for a laser transition between the two manifolds [(S,L)J] and [(S',L')J'] can be calculated using the relation [28]:

$$\sigma\left(\lambda_p\right) = \frac{\lambda_p^4}{8\pi c n^2 \Delta\lambda_{eff}} A_{rad}\left[(S,L)J;(S',L')J'\right] \tag{13}$$

where λ_p is the peak wavelength of the emission line and $\Delta\lambda_{eff}$ is the effective fluorescence line-width determined by integrating the fluorescence line shape and dividing it by the intensity at λ_p [29, 33]. The effective line-width is used as 'width of the line' for discussion pertaining to the asymmetric nature of emission line.

However, the integrated emission cross-section of a transition between the manifolds [(S,L)J] and [(S',L')J'] is given by the expression [29]:

$$\int \sigma\left(\bar{v}\right)d\bar{v} = \frac{\lambda_p^2}{8\pi c n^2} A_{rad}\left[(S,L)J;(S',L')J'\right] \tag{14}$$

This expression is useful if information about line width is not available.

4.5. Quantum Efficiency

The quantum efficiency, η, of the metastable level manifold [(S,L)J] is defined by the expression [29]:

$$\eta = \frac{\tau_f}{\tau_r} \tag{15}$$

where τ_f is the measured fluorescence lifetime of the metastable level and τ_r is the calculated radiative lifetime of this level. In the present work, measurement on τ_f has not been carried out therefore quantum efficiency is not reported.

5. HYPERSENSITIVE TRANSITION

The intensities of the *f-f* transitions of trivalent lanthanides are, in general, marginally affected by the environment of the ion in a host matrix. However, a few transitions are very sensitive to the environment and are usually more intense when the ion is incorporated in complexes rather than when the ion is in aqueous solution. These are called hypersensitive transitions [34], and they obey the selection rules $|\Delta J| \leq 2$, $|\Delta L| \leq 2$ and $\Delta S = 0$. Judd [15] pointed out that among the three intensity parameters; Ω_2 is most affected by changes in the environment around the rare earth ion in the host matrix. The hypersensitivity was strongly correlated with those transitions having large matrix elements of $U^{(2)}$. Jorgensen *et al.* [35] examined some of the potential mechanism for hypersensitive transitions and suggested that the origin of these transitions is due to non-uniform electric field across the ion, which in turn arises because the dipoles are induced in the surrounding of the ion in inhomogeneous dielectric due to the radiation field. Judd [36] pointed out that the symmetry arguments could be introduced to classify site symmetries at rare earth ion which might promote hypersensitivity. Peacock [37] has reviewed some of the arguments including vibronic mechanism, covalency, pure quadrupole radiation, inhomogeneous dielectric, dynamic coupling mechanism, *etc.* and discussed hypersensitivity in terms of a correlation between oscillator strength and ligand basicity. However, the problem of hypersensitivity is yet not fully resolved.

6. ION-ION INTERACTION

The ion-ion interaction between two similar rare earth ions can cause the fluorescence quenching and hence a reduction in the radiative lifetime of the metastable laser level [38]. The fluorescence quenching primarily depends upon the concentration of rare earth ions in a given host matrix. Higher the concentration of rare earth ions, the smaller will be the ion-ion separation and consequently larger will be the ion-ion interaction resulting in a higher fluorescence quenching. This quenching mechanism is also known as self-quenching or the concentration quenching. The self quenching of rare earth ions limits the optical pumping efficiency, which would have been, otherwise, possible to be enhanced by increasing the concentration of the ions. Fig. **2** shows the self-quenching between a pair of Nd^{3+} ions. The excited neodymium ion in the metastable state $^4F_{3/2}$ can decay non radiatively to the $^4I_{15/2}$ state and, in turn, excite a nearby Nd^{3+} ion from the state $^4I_{9/2}$ to the state $^4I_{15/2}$; which subsequently decay back to the state $^4I_{9/2}$ *via* non radiative transitions. This mechanism thus leads to the fluorescence quenching in the former ion. The fluorescence quenching increases as the average Nd^{3+} ion-ion distance decreases with increasing neodymium ion concentration in the host.

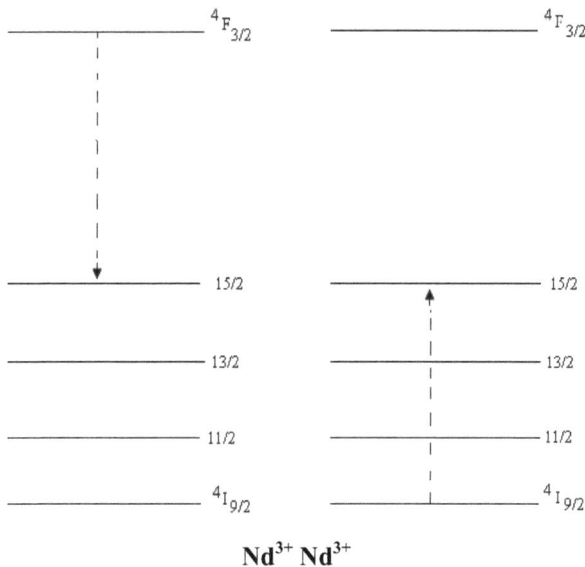

Figure 2: Self Quenching of $^4F_{3/2}$ Level in Nd^{3+} ion.

The self-quenching is also dependent on the type of the host matrix [38], since the separation between rare earth ions may be different in different host matrices.

7. PHYSICAL CHARACTERIZATIONS

7.1. Density and Molar Volume

The value of density (D) is used to determine the concentration of rare earth ions and molar volume (V_M) of the prepared glass samples. The concentration of rare earth ions is of importance in computing the experimental value of the oscillator strength of any electronic transitions and is also useful in the determination of some other physical parameters of the samples *viz.*, inter-ionic separation and ionic radius. Molar volume and density being sensitive to the structure of the system [39]; their measurements were of help in understanding the structure of the prepared glasses. The value of D and V_M is displayed in Table **1**. The concentration of rare earth ions in the samples was calculated using the expression [40]

$$C = \frac{D \times R_n}{M} \times 1000 \qquad (16)$$

$$N = \frac{D \times R_n}{M} \times N_A \qquad (17)$$

where C is the concentration in moles/1000 cm^3, N is the concentration in number of ions per cm^3, R_n the number of moles of rare earth ions and N_A is the Avogadro's number (= 6.0248×10^{23}). The calculated values of N for all the prepared glasses are included in Table **1**. Further, the calculated values of inter-ionic separation (r_i) and ionic radius (r_p) for all the glass samples are also displayed in Table **1**.

7.1.1. 20MO·xBi₂O₃·(79.5-x)B₂O₃·0.5Er₂O₃ (M = Zn and Cd) Glasses

It is seen from Table **1** that the density depends upon the type of transition metal ion as well as heavy metal modifiers, *i.e.* unconventional network formers (Bi$_2$O$_3$), which are known to occupy the interstitial positions in the glassy positions. From this table, it is observed that the samples containing CdO have higher density than those containing ZnO. Because the Cd^{2+} ions are heavier than the Zn^{2+} ions, therefore higher value of density is expected in CBBE glasses than

ZBBE glasses. The replacement of B_2O_3 (atomic mass 61.83 g) by Bi_2O_3 (atomic mass 465.95 g) results in an increase in the density of the above glasses.

7.1.2. $20MO \cdot xSiO_2 \cdot (79.5-x)Bi_2O_3 \cdot 0.5Er_2O_3$ (M = Zn and Cd) Glasses

In case of ZSBE and CSBE glasses, density as well as molar volume decrease with decrease in Bi_2O_3 content or increase in SiO_2 content because Bi_2O_3 (atomic mass 496.95) is heavier than SiO_2 (atomic mass 60.09). The molar volume behaviour reveals that addition of SiO_2 contracted the structure of the loose network in these glasses. The smaller values of radii [41] and bond length [42] of SiO_2 than that of Bi_2O_3 results in a shrinking of the free volume which decreases the overall molar volume of these glasses. This trend supports the idea of shrinking-structure, which may be attributed to the ability of the voids of the bismuthate network to accommodate such modifier ions without any expansion of the glass matrix. Further, according to Pan *et al.* [43], SiO_2 tetrahedra can be incorporated in the network of more flexible structure of bismuth oxygen polyhedra, $[BiO_n]$ due to more ionic nature of the Bi-O bond and hence there is decrease in the molar volume in the present glasses (ZSBE and CSBE glasses). Other physical parameters, *viz.*, number density of Er^{3+} ions, inter-ionic distance and ionic radius of all the glasses has been estimated and presented in Table **1**.

8 OPTICAL CHARACTERIZATIONS

8.1. Refractive Index

The values of refractive index obtained with the help of Brewster angle method are listed in Table **1** for all the ZBBE, CBBE, ZSBE and CSBE glasses. Similar results were obtained by Saisudha *et al.* [13] and Chen *et al.* [18] in binary bismuth borate glasses doped with different rare earth ions. In the present glasses, the change in refractive index is in accordance with change in concentration of heavy metal oxide, *i.e.* Bi_2O_3.

The glass sample CSBE1 has the highest value of refractive index (2.22) among all the prepared glasses. In general, the glasses with higher Bi_2O_3 content have higher value of *n* within a given glass composition. A close examination of the variations in *n* and *D* values with composition reveals that the changes in refractive index and density are concurrent, *i.e.*, if density increases/decreases, the refractive index also increases/decreases. Higher density implies a more compact structure and the compact structure leads to a higher value of refractive index [44]. From the value of n, dielectric constant (ε) and reflection loss (R_L) have also been calculated and their values are presented in Tables **1** for all the glass samples.

Table 1: Density (D), molar volume (V_M), number density of Er^{3+} ions (N), inter-ionic distance (r_i), ionic radius (r_p), refractive index (n), dielectric constant (ε) and reflection loss (R_L), for Er^{3+} doped ZBBE, CBBE, ZSBE and CSBE glasses

Sample code	x (mol%)	D (g/cm³)	V_M (cm³/mol)	N (10^{20} ions/cm³)	r_i (Å)	r_p (Å)	n	ε (n²)	R_L (%) $[(n-1)/(n+1)]^2$
ZBBE1	**15**	**4.39**	**29.29**	**2.35**	**7.52**	**3.07**	**1.75**	**2.95**	**7.43**
ZBBE2	20	4.81	30.82	2.31	7.56	3.08	1.82	3.20	8.45
ZBBE3	25	5.19	32.42	2.21	7.67	3.09	1.88	3.38	9.34
ZBBE4	30	5.39	34.99	2.11	7.79	3.14	1.92	3.61	9.93
ZBBE5	35	5.59	37.35	2.03	7.89	3.18	1.95	3.96	10.37
CBBE1	**15**	**4.35**	**31.56**	**3.82**	**6.39**	**2.58**	**1.74**	**3.03**	**7.29**
CBBE2	20	4.67	33.77	3.57	6.55	2.64	1.79	3.20	8.02
CBBE3	25	5.03	35.35	3.41	6.64	2.68	1.86	3.46	9.04
CBBE4	30	5.18	38.21	3.15	6.82	2.75	1.88	3.53	9.34
CBBE5	35	5.37	40.62	2.97	6.96	2.80	1.91	3.65	9.78
ZSBE1	**10**	**6.99**	**49.77**	**1.82**	**8.19**	**3.30**	**2.19**	**4.79**	**13.92**
ZSBE2	20	6.89	44.61	2.03	7.89	3.18	2.17	4.71	13.62
ZSBE3	30	6.63	40.28	2.24	7.64	3.08	2.13	4.53	13.03
ZSBE4	40	6.44	35.13	2.57	7.30	2.94	2.09	4.37	12.44
ZSBE5	50	6.29	29.52	3.06	6.88	2.76	2.06	4.24	11.99
CSBE1	**10**	**7.19**	**49.69**	**1.81**	**8.19**	**3.30**	**2.22**	**4.92**	**14.35**
CSBE2	20	7.08	44.74	2.01	7.91	3.18	2.20	4.84	14.06
CSBE3	30	7.01	39.43	2.29	7.58	3.05	2.19	4.79	13.91
CSBE4	40	6.94	33.95	2.66	7.22	2.91	2.18	4.75	13.76
CSBE5	50	6.68	29.17	3.09	6.86	2.76	2.13	4.53	13.03

8.2 Optical Absorption

8.2.1. (A) $20MO \cdot xBi_2O_3 \cdot (79.5-x)B_2O_3 \cdot 0.5Er_2O_3$ (M = Zn and Cd) Glasses

Er^{3+}-doped glass is one of the most important laser material because the 1.5 μm emission from the $^4I_{13/2} \rightarrow {}^4I_{15/2}$ transition is eye safe and is located in the optical communication window. In order to meet the demand of the exponential increase of information transmission, the Er^{3+}-doped fiber amplifier is the key for the communication systems. Therefore, Er^{3+}-doped metal oxide glasses especially bismuth oxide [5] have been proved to be promising candidates as gain medium for the 1.5 μm fiber amplifier because of their high refractive index, high transparency for infrared light, large glass forming region and good chemical solubility of the these ions in the host. The observed optical absorption spectra of Er^{3+}-doped bismuth borate glasses are shown in Figs. **3** and **4** for ZBBE and CBBE glasses, respectively. All the spectra consist of number of absorption bands that arise due to the *f-f* electronic transition from the ground states manifold $^4I_{15/2}$ to the manifolds of various excited states of Er^{3+} ions. On the basis of earlier reports [45-47], the assignment for electronic transitions has been made. The observed band positions of various electronic transitions, in the wavelength range 420-1600 nm, are presented in Table **2**. From Figs. **3** and **4** it is observed that the spectra are similar in shape only and there is change in band position for the particular transition.

8.2.2. 20MO·xSiO₂·(79.5-x)Bi₂O₃·0.5Er₂O₃ (M = Zn and Cd) Glasses

The optical absorption spectra for Er^{3+} doped zinc/cadmium bismuth silicate glasses are shown in Figs. **5** and **6**, respectively. The observed spectra consist of many transitions from the ground state $^4I_{15/2}$. The observed band positions in ZSBE and CSBE glasses are presented in Table **3**. The observed spectra are similar in shape but the relative intensities of the bands vary with the host glass composition. It is also noted that there is marginal shift in the band positions with change in host glass compositions. Due to strong absorption of host glasses at the ultra-violet range, the absorption bands at wavelength shorter than 450 nm could not be distinguished. The experimental oscillator strength (f_{expt}) is determined with the help of Eq. (2), and is used to calculate the theoretical oscillator strength with least square analysis method. The values of experimental and theoretically calculated oscillator strength and r.m.s. deviation are presented in Tables **2** and **3** for all the prepared glass systems, respectively. The theoretically calculated values of oscillator strength, f_{cal}, and the Judd-Ofelt intensity parameters, Ω_λ, are calculated, utilizing the matrix elements $\|U^{(\lambda)}\|^2$. The r.m.s. deviations obtained are also presented in Tables 2 and 3, respectively. The low values of r.m.s. deviation as judged by the values reported by other workers [45, 48-51] signify a good agreement between f_{expt} and f_{cal}. The values of intensity parameters Ω_2, Ω_4 and Ω_6 are listed in Table **4** for all the samples.

Out of the three Judd-Ofelt intensity parameters, Ω_2 is known to be most sensitive to local structural changes around rare earth ions in the host [18] and is related to the covalency to the RE-O bond. Ω_2-intensity parameter shows unique variation with Bi_2O_3 content in the host glasses. As the Bi_2O_3 content increases, Ω_2 parameter increases in ZBBE and CBBE glasses but in case of ZSBE and CSBE glasses it decreases. A similar trend is also observed by Izumitani *et al.* [49] in silicate and phosphate glasses. The ratio, Ω_4/Ω_6, known as spectroscopic quality factor (SQF), is found to be greater than one in the present study. The ratio Ω_4/Ω_6 in the borate glasses is 1.71, and in the present glasses it is 1.73 (CSBE1 glass sample). This indicates that present glasses are fairly rigid. Nageno *et al.* [50] have also studied various alkali borate and phosphate glasses and reached to similar conclusion that Ω_4 and Ω_6 parameters are related to the rigidity of the glass matrix. It is further seen from Table **4** that the parameters Ω_4 and Ω_6 also vary with composition of the glass; these variations are responsible for varying radiative properties of Er^{3+} ion in the prepared glasses. On the basis of information obtained from the fluorescence measurements along with the calculated values of the intensity parameters Ω_4 and Ω_6, the optical properties of Er^{3+} ions are evaluated.

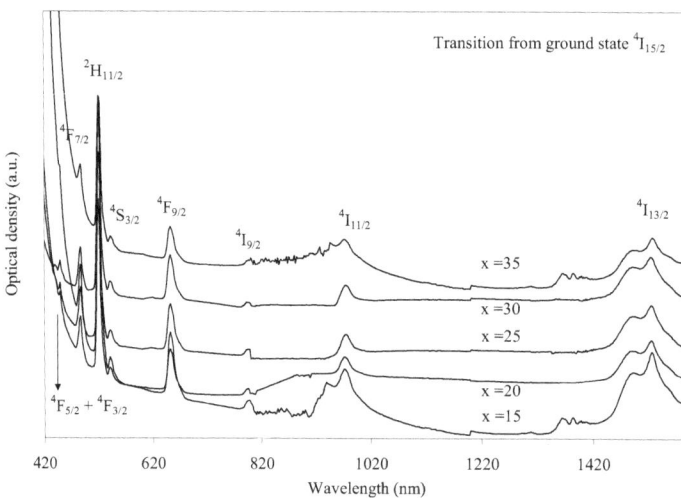

Figure 3: Optical absorption spectra of $20ZnO \cdot xBi_2O_3 \cdot (79.5-x)B_2O_3 \cdot 0.5Er_2O_3$ (ZBBE) glasses.

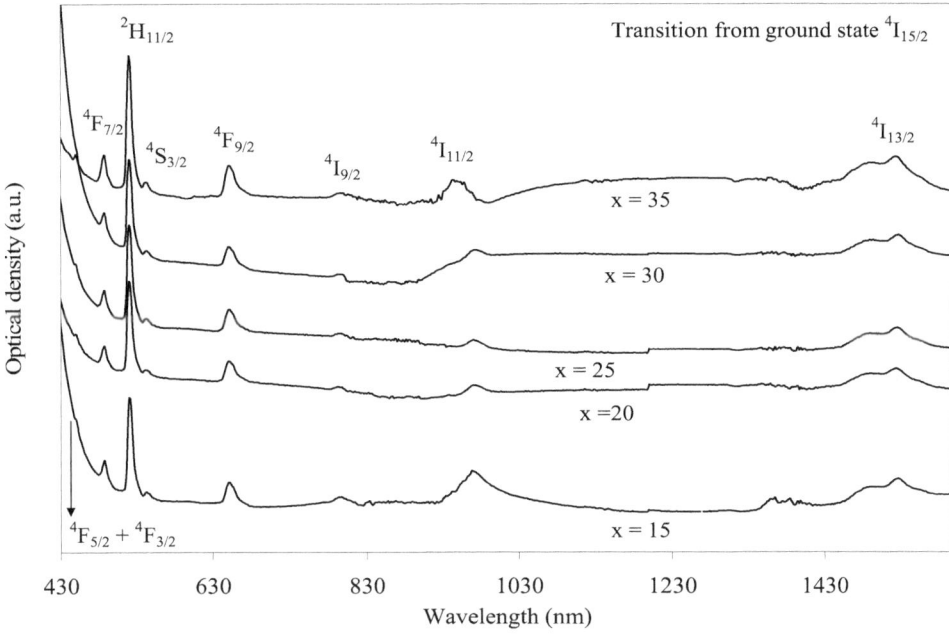

Figure 4: Optical absorption spectra of $20CdO \cdot xBi_2O_3 \cdot (79.5-x)B_2O_3 \cdot 0.5Er_2O_3$ (CBBE) glasses.

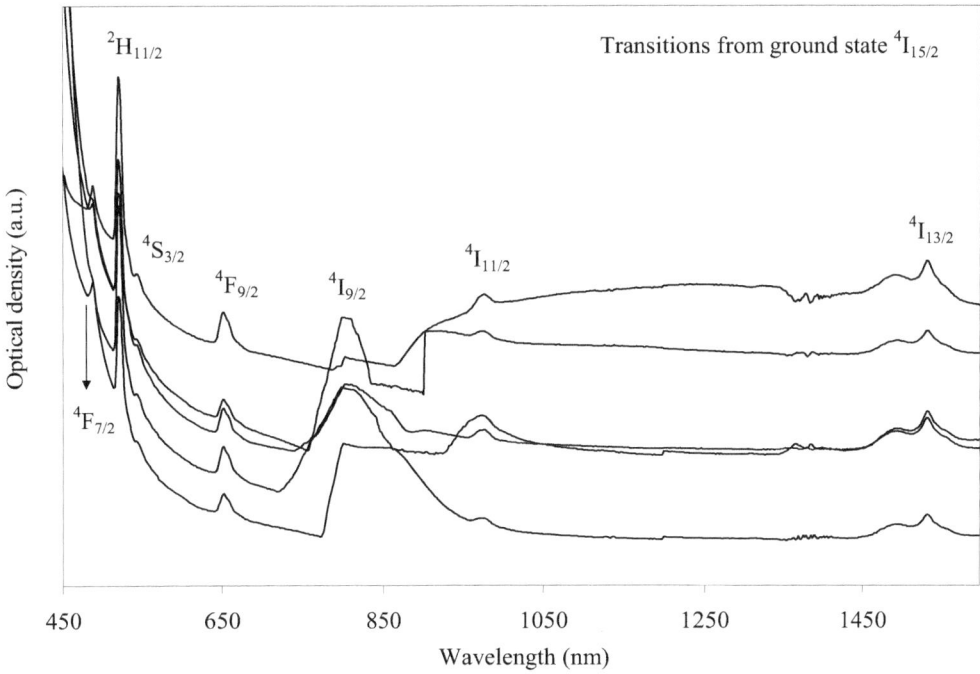

Figure 5: Optical absorption spectra of $20ZnO \cdot xSiO_2 \cdot (79.5-x)Bi_2O_3 \cdot 0.5Er_2O_3$ (ZSBE) glasses.

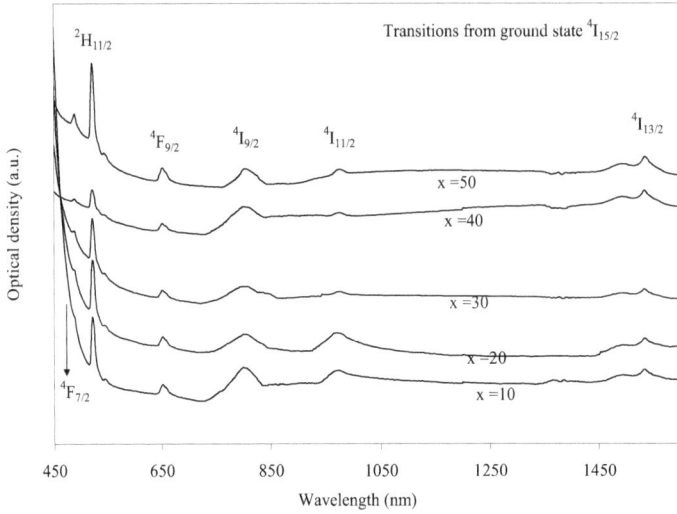

Figure 6: Optical absorption spectra of 20CdO·xSiO$_2$·(79.5-x)Bi$_2$O$_3$·0.5Er$_2$O$_3$ (CSBE) glasses.

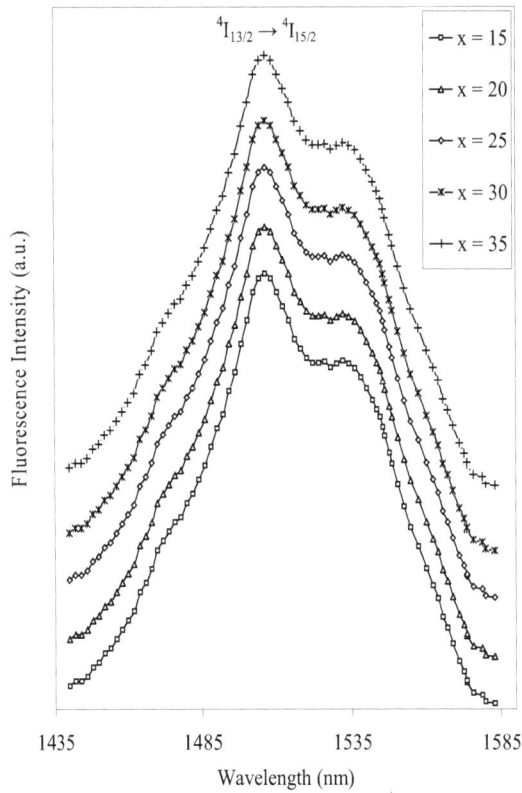

Figure 7: Emission spectra of 20ZnO·xBi$_2$O$_3$·(79.5-x)B$_2$O$_3$·0.5Er$_2$O$_3$ (ZBBE) glasses.

8.3 Fluorescence Spectra

8.3.1. 20MO·xBi$_2$O$_3$·(79.5-x)B$_2$O$_3$·0.5Er$_2$O$_3$ (M = Zn and Cd) Glasses

The fluorescence spectra of Er^{3+} ions doped ZBBE and CBBE glasses recorded at RT are shown in Figs. **7** and **8**, respectively. The fluorescence spectra are recorded in the range 1050-1650 nm for the excitation wavelength, λ_{exc}= 990 nm. Each spectrum exhibited the transition $^4I_{13/2} \rightarrow {}^4I_{15/2}$ at 1.5 μm in the prepared glasses. The fluorescence spectra obtained in the present study are qualitatively similar as observed by Chen *et al.* [18] and Oprea *et al.* [54]. Also from Figs. **7** and **8**, it is observed that the fluorescence intensity of the different transitions is dependent on the composition of the host glasses.

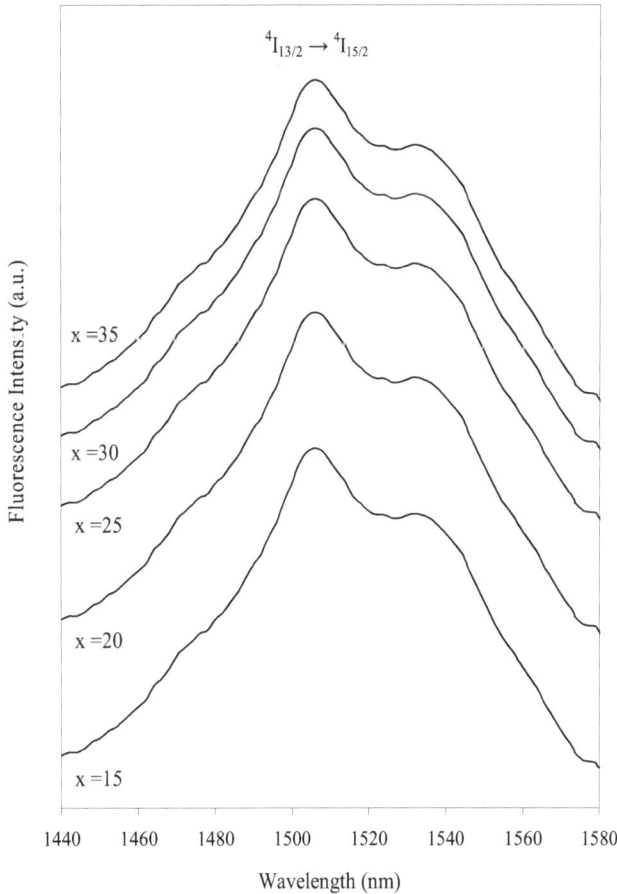

Figure 8: Emission spectra of 20CdO·xBi$_2$O$_3$·(79.5-x)B$_2$O$_3$·0.5Er$_2$O$_3$ (CBBE) glasses.

8.3.2. 20MO·xSiO₂·(79.5-x)Bi₂O₃·0.5Er₂O₃ (M = Zn and Cd) Glasses

The fluorescence spectra of Er^{3+} ions doped ZSBE and CSBE glasses recorded in the range 1050-1650 nm for excitation wavelength $\lambda_{exc.}=$ 990 nm are shown in Figs. **9** and **10**, respectively. The radiative properties of Er^{3+} ions in glasses play significant role in deciding their laser performance, it is important to optimize these properties by selecting the host glass composition. This section presents the interpretation of the radiative properties of Er^{3+} ions obtained in the prepared glasses.

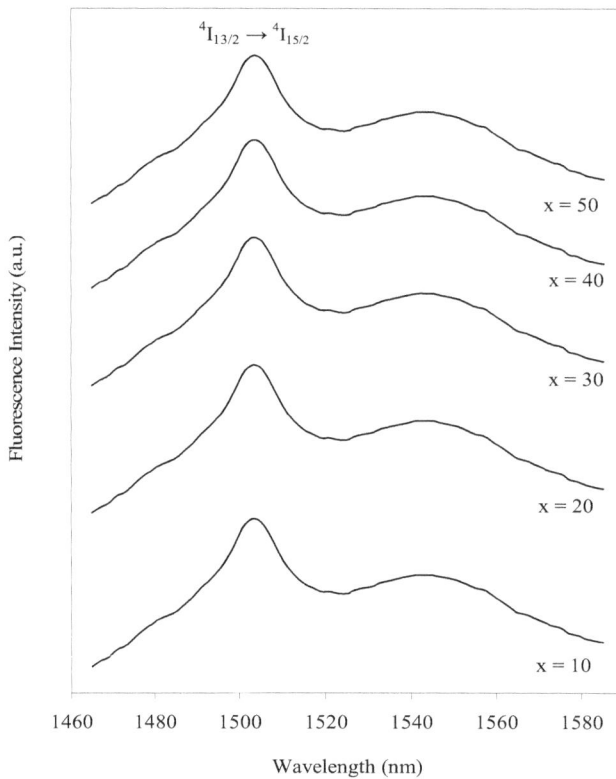

Figure 9: Emission spectra of 20ZnO·xSiO₂·(79.5-x)Bi₂O₃·0.5Er₂O₃ (ZSBE) glasses.

The radiative property related to $^4I_{13/2} \rightarrow {}^4I_{15/2}$ transition is discussed in detail because the emission at 1.5 μm lies in the third communications window where the losses are minimum. By using the peak wavelength of the emission band, λ_p, the effective bandwidth $\Delta\lambda_{eff}$, and Judd-Ofelt intensity parameters (Ω_2, Ω_4, and Ω_6); the radiative properties of Er^{3+} ion in the prepared bismuth borate/silicate glasses are calculated. These are presented and discussed in the next section.

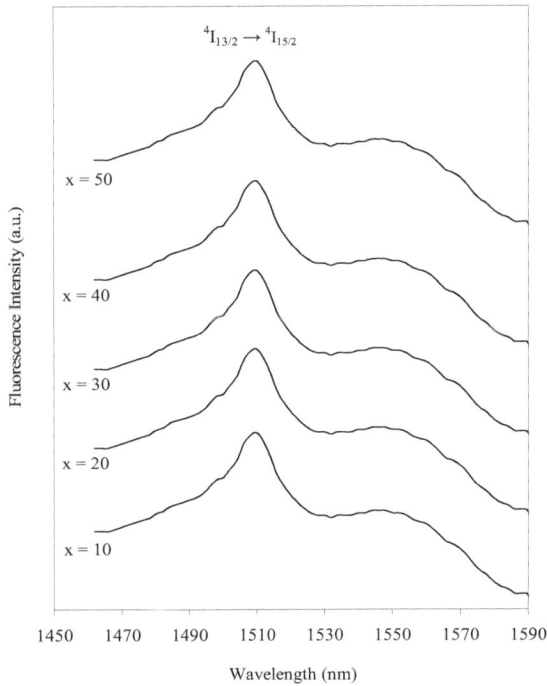

Figure 10: Emission spectra of $20CdO \cdot xSiO_2 \cdot (79.5-x)Bi_2O_3 \cdot 0.5Er_2O_3$ (CSBE) glasses.

8.4 Radiative Properties

The radiative properties of Er^{3+} ions in glasses are of great importance in determining their suitability for producing laser devices and also for other optical applications such as in solar concentrators, optical detector, waveguide laser, optical fibers, fluorescent display devices, bulk laser, IR to visible upconverters, phosphors *etc.* The emission of Erbium at 1.5 µm ($^{4}I_{3/2} \rightarrow {}^{4}I_{15/2}$ transition) is eye-safe. The radiative properties discussed are spontaneous emission probability, radiative life time, stimulated emission cross-section and figure of merit.

8.4.1. Spontaneous Emission Probability

The spontaneous emission probability, A_{rad} (s^{-1}), for an electric dipole transition from metastable level manifold [(S,L)J] to the lower lying state manifold [(S',L')J'] is calculated using Eq. (10). For Er^{3+} ion, the metastable level is $^{4}I_{13/2}$ and the lower level to which transition terminates is $^{4}I_{15/2}$. The obtained values of A_{rad} are presented in Table **5** for all the glass samples under study. It is observed from the tables that the values of transition probability (A_{rad}) increase with increase in Bi_2O_3 content in ZBBE and ZSBE glasses whereas in CBBE and CSBE glasses the behaviour is reverse. Such variations in A_{rad} can be explained as suggested by Izumitani *et al.* [49] for silicate glasses. Since similar results were obtained for ZBBS, CBBS, ZSBS and CSBS glasses [21].

Table 2: Oscillator strength for some transitions from the ground level $^4I_{15/2}$ to the indicated levels and their root mean square deviation (∂_{RMS}), which indicates the fit quality of theoretical and experimental results for Er^{3+} doped ZBBE and CBBE glasses

Transitions from ground level	λ (nm)	Oscillator strength (10^{-8})									
		ZBBE1		ZBBE2		ZBBE3		ZBBE4		ZBBE5	
$^4I_{15/2} \rightarrow$	f_{expt}	f_{cal}	f_{expt}	f_{cal}	f_{expt}	f_{cal}	f_{expt}		f_{cal}	f_{expt}	f_{cal}
$^4I_{13/2}$	1522	245	239	251	238	257	242	263	249	268	251
$^4I_{11/2}$	972	68	76	72	65	72	87	83	87	92	93
$^4I_{9/2}$	790	39	41	36	42	44	48	55	49	63	73
$^4F_{9/2}$	650	234	239	242	248	257	263	284	298	323	320
$^4S_{3/2}$	544	35	54	41	58	43	59	52	63	50	61
$^2H_{11/2}$	520	801	822	861	869	899	934	969	972	1043	1045
$^4F_{7/2}$	488	190	211	209	231	234	264	269	284	274	297
$^4F_{5/2}+^4F_{3/2}$	450	39	49	58	64	46	55	25	38	31	33
$\partial_{RMS}(10^{-7})$		3.74		4.06		4.73		5.27		4.64	
		CBBE1		CBBE2		CBBE3		CBBE4		CBBE5	
$^4I_{13/2}$	1522	506	443	535	505	573	554	596	597	591	589
$^4I_{11/2}$	966	83	77	129	159	199	191	293	280	239	299
$^4I_{9/2}$	792	311	328	352	298	401	424	434	441	454	483
$^4F_{9/2}$	650	763	764	799	795	824	837	897	888	836	832
$^4S_{3/2}$	540	157	173	260	254	261	268	275	281	311	351
$^2H_{11/2}$	520	270	279	349	350	358	348	371	385	376	388
$^4F_{7/2}$	488	169	153	220	315	225	320	279	277	271	278
$^4F_{5/2}+^4F_{3/2}$	450	346	330	—		—		—		—	—
$\partial_{RMS}(10^{-7})$		2.78		1.17		1.32		1.37		0.97	

Also, A_{rad} value is higher for glasses containing ZnO than for CdO. It is suggested that the value of A_{rad} is controlled by the crystal field parameter, A_{tp}, which is related to the asymmetry in ligand field around the rare earth ions. Asymmetry is primarily influenced by the ionic field strength (Ze/a^2) of the modifier cation present in the glass structure. Higher the value of ionic strength, larger is the distortion in the cage surrounding rare earth ion and consequently higher will be the parameter A_{tp}. The ionic strength of Zn^{2+} ions is significantly higher than that of Cd^{2+} ions, therefore, the values of A_{rad} are expected to be higher for ZBBE and ZSBE glasses than those for CBBE and CSBE glasses, respectively. Since the ionic field strength increases with Bi_2O_3 content in the host glasses, therefore, A_{rad} also increases in the present glasses. The spontaneous emission probabilities $A_{JJ'}$ and A_{rad} for Er^{3+} ions doped presently investigated glasses (ZBBE, CBBE, ZSBE and CSBE) have the same order as reported by Canalejo *et al.* [52]. It is thus inferred that the modifiers are responsible for the increased distortion in the cage around the rare earth ions by lowering the site symmetry (or increase in site asymmetry). This leads to an increase in the transition probability. The types of anion present in the glass also affect the transition probabilities.

Table 3: Oscillator strength for some transitions from the ground level $^4I_{15/2}$ to the indicated levels and their root mean square deviation (∂_{RMS}), which indicates the fit quality of theoretical and experimental results for Er^{3+} doped ZSBE and CSBE glasses

Transitions from ground level	λ (nm)	Oscillator strength (10^{-8})									
		ZSBE1		ZSBE2		ZSBE3		ZSBE4		ZSBE5	
$^4I_{15/2} \rightarrow$		f_{expt}	f_{cal}	f_{expt}	f_{cal}	f_{expt}	f_{cal}	f_{expt}	f_{cal}	f_{expt}	f_{cal}
$^4I_{13/2}$	1532	221	217	212	205	201	187	175	152	104	97
$^4I_{11/2}$	972	204	237	149	128	127	122	124	99	139	154
$^4I_{9/2}$	802	312	366	308	318	299	306	269	209	255	213
$^4F_{9/2}$	652	291	307	242	291	222	289	184	254	127	198
$^2H_{11/2}$	520	398	399	383	384	333	343	299	301	223	225
$^4F_{7/2}$	488	----	----	----	----	-----	----	559	568	392	349
$\partial_{RMS}(10^{-6})$		2.31		2.00		1.89		3.06		2.82	
		CSBE1		CSBE2		CSBE3		CSBE4		CSBE5	
$^4I_{13/2}$	1532	229	269	174	189	96	137	86	109	69	73
$^4I_{11/2}$	970	358	324	229	252	226	274	151	158	64	74
$^4I_{9/2}$	802	442	374	337	415	228	307	203	242	197	132
$^4F_{9/2}$	652	224	322	207	281	180	223	154	205	123	105
$^2H_{11/2}$	520	368	373	326	329	302	305	277	278	219	217
$^4F_{7/2}$	488	----	----	----	----	----	-----	391	298	378	355
$\partial_{RMS}(10^{-6})$		3.38		2.38		1.89		1.46		1.44	

Table 4: Judd-Ofelt intensity parameters (Ω_2, Ω_4, Ω_6) of Er^{3+} ions doped ZBBE, CBBE, ZSBE and CSBE glasses

Glass	Ω_2 (10^{-20}cm^2)	Ω_4 (10^{-20}cm^2)	Ω_6 (10^{-20}cm^2)	SQF (Ω_4/Ω_6)
ZBBE1	**2.12**	**1.56**	**1.44**	**1.08**
ZBBE2	2.83	1.39	1.09	1.27
ZBBE3	3.02	1.32	1.22	1.08
ZBBE4	3.69	1.21	1.11	1.09
ZBBE5	3.94	1.48	1.07	1.38
CBBE1	**3.05**	**2.84**	**2.17**	**1.31**
CBBE2	3.08	1.77	1.61	1.09
CBBE3	3.14	1.78	1.56	1.14
CBBE4	3.35	2.35	1.66	1.41
CBBE5	3.45	1.44	1.05	1.37
ZSBE1	**0.98**	**1.01**	**0.85**	**1.18**
ZSBE2	0.78	1.15	0.91	1.26
ZSBE3	0.65	1.21	1.07	1.13
ZSBE4	0.58	1.97	1.22	1.61
ZSBE5	0.35	2.14	2.13	1.01
CSBE1	**0.92**	**1.72**	**0.99**	**1.73**
CSBE2	0.72	1.69	1.15	1.47
CSBE3	0.70	1.48	1.24	1.19
CSBE4	0.60	1.38	1.37	1.01
CSBE5	0.55	1.23	1.67	0.74
$Er_2P_5O_{14}$	1.88	1.34	1.13	1.19
ZBLAN	2.91	1.27	1.11	1.14
S20AN	6.64	1.10	0.41	2.68
S25AN	9.53	1.49	0.83	1.79

Therefore, the transition probabilities of emission transitions of Er^{3+} ions can be varied by varying Bi_2O_3 content in the host. The largest value of A_{rad} for transition $^4I_{13/2} \rightarrow {}^4I_{15/2}$ is responsible for maximum fluorescence intensity.

8.4.2 Radiative Lifetime

The lifetime of the metastable level $^4I_{13/2}$ is an important parameter in determining the energy storage in Er^{3+}-doped glass lasers. The radiative lifetime, τ_r of the level $^4I_{13/2}$, which decays radiatively to lower level $^4I_{15/2}$ is calculated from radiative decay probability of level $^4I_{13/2}$ using Eq. (11). The values of τ_r for present glasses are included in Table **5**. The radiative lifetime depends upon the transition probabilities (Eq. (11), which in turn depends upon the non-centrosymmetric local field at the rare earth ion. The radiative lifetime is a good indicator of the strength of this non-centrosymmetric field. The strength of this local field is determined by the distortion in the ligands (or the cage) surrounding the rare earth ions. Higher the distortion, higher is the transition probability and smaller will be the radiative lifetime. For a fixed glass former cation, the distortion in the ligands may be altered by the following number of factors, *i.e.*

a) **Field strength of the modifier cations;** higher the field strength of the modifier cation, larger will be the distortion in the cage around the rare earth ion.

b) **Field strength of the anions;** the anions of higher field strengths are expected to enhance the cage distortion.

c) **Size of the modifier cations and those of the interstitial sites occupied by them;** larger the differences between them, greater would be the cage distortion.

d) **Particular interstitial sites occupied by the modifier cation;** if the cation goes to first nearest interstitial site to rare earth ion, the resulting distortion in the cage would be large.

Besides these, the radiative lifetime also depends upon ion-ion interaction (between rare earth ions). Higher the ion-ion interaction, higher will be the self quenching and, therefore, smaller will be the radiative lifetime [38]. The ion-ion interaction and, hence the radiative lifetime, depends upon

a) **Concentration of rare earth ions;** higher the concentration of rare earth ions in glass, larger will be the ion-ion interaction, and

b) **Size of modifier cations;** modifier cations of larger size are expected to increase the ion-ion separation and therefore lead to a decrease in the ion-ion interaction.

In the present study, the ZBBE and ZSBE glasses contain Zn^{2+} as primary modifier cations which have a higher value of field strength as compared to that of Cd^{2+} modifier cations present in CBBE and CSBE glasses. Also, Zn^{2+} ions are smaller than the Cd^{2+} ions. Therefore, factors (a) and (f) would lead to smaller values of τ_r for ZBBE and ZSBE glasses than those for CBBE and CSBE glasses, respectively. The radiative lifetime varies with Bi_2O_3 content in the host glass as seen from Table **5**. Therefore, from the present study it is concluded that the energy storage efficiency in cadmium bismuth borate/silicate glasses is more than that in the zinc bismuth borate/silicate glasses.

Table 5: The peak wavelength (λ_p), radiative transition probability (A_{rad}), radiative life time (τ_r), stimulated emission cross-section (σ), line width ($\Delta\lambda_{eff}$), figure of merit ($\sigma \times \Delta\lambda_{eff}$), and the total emission cross-section (σ_t) for $^4I_{13/2} \rightarrow {}^4I_{15/2}$ transition in Er^{3+} doped ZBBE, CBBE, ZSBE and CSBE glasses

Sample No.	Transitions from $^4I_{13/2} \rightarrow {}^4I_{15/2}$					
	λ_p (nm)	A_{rad} (s^{-1})	τ_r (ms)	σ (10^{-21}cm^2)	$\Delta\lambda_{eff}$ (nm)	FOM ($\sigma \times \Delta\lambda_{eff}$)
ZBBE1	**1506**	**1934**	**0.517**	**4.34**	**66**	**286.44**
ZBBE2	1510	1699	0.588	3.39	69	233.91
ZBBE3	1510	2080	0.481	3.84	70	268.80
ZBBE4	1515	2023	0.494	3.72	68	252.96
ZBBE5	1519	2076	0.482	3.73	68	253.64
σ_t				19.02		
CBBE1	**1506**	**362**	**2.762**	**8.36**	**76**	**635.36**
CBBE2	1509	254	3.937	7.79	73	545.30
CBBE3	1510	283	3.533	7.61	71	570.75
CBBE4	1510	294	3.401	7.97	71	573.84
CBBE5	1515	271	3.690	7.64	68	519.52
σ_t				39.37		
ZSBE1	**1510**	**947**	**1.055**	**1.95**	**70**	**136.50**
ZSBE2	1508	897	1.115	1.87	70	130.90
ZSBE3	1505	858	1.165	1.95	66	128.70
ZSBE4	1505	761	1.314	1.94	61	118.34
ZSBE5	1500	709	1.410	1.84	61	112.24
σ_t				9.55		
CSBE1	**1516**	**176**	**5.681**	**0.46**	**54**	**24.84**
CSBE2	1511	195	5.128	0.55	51	28.05
CSBE3	1510	223	4.484	0.67	48	32.16
CSBE4	1506	246	4.065	0.74	48	35.52
CSBE5	1501	258	3.876	0.85	45	38.25
σ_t				3.27		
Silicate				0.65	40	26.00
Phosphate				0.86	46	39.56
Tellurite				0.85	66	56.10
PbO-PbF$_2$-B$_2$O$_3$				0.50	60	30.00

8.4.3. Stimulated Emission Cross-Section

The stimulated emission cross-section is an important parameter in determining the energy extraction efficiency of a lasing transition. The fluorescence transitions with large stimulated emission cross-sections are, in general, characterized as good lasing transitions. The peak stimulated emission cross-section, σ (cm^2), of a fluorescent transition can be computed using Eq. (13). It can be predicted from this equation that the stimulated emission cross-section depends upon

a) The transition probabilities and hence on the Judd-Ofelt intensity parameters, Ω_λ.

b) The emission bandwidth, $\Delta\lambda_{\text{eff}}$.

c) The refractive index, n, of the host material.

d) The Ω-parameters dependence on host compositions of prepared glass samples.

Larger the asymmetry around rare earth ions, higher will be the value of Ω_λ parameters [26, 49] and hence higher the value of stimulated emission cross-section. It is important to mention that the stimulated emission cross-section of a transition can be enhanced by suitably varying the composition of the host glass so as to result in a high value of the particular Ω_λ which corresponds to the maximum value of the corresponding matrix element $\|U^{(\lambda)}\|^2$ of the transition of interest.

Emission bandwidth is a measure of the extent of the Stark splitting in the initial and final manifolds and is inhomogeneous due to the site-to-site variations in the local field seen by rare earth ion in the glass. Glasses, in general, have multiplicity of sites. Lesser the number of sites available for rare earth ions in a glass, the smaller will be the inhomogeneous broading. Smaller anionic field strength also results in smaller Stark splitting and hence the narrower linewidth [53]. The linewidth usually increases with the increasing charge and decreasing size of modifier cations [54]. Smaller values of $\Delta\lambda_{eff}$ lead to higher σ, (Eq. (13)). Since the transition probability A_{rad} depends upon the host refractive index n as

$[n(n^2+2)^2/9]$, (Eq. (10)), the stimulated emission cross-section depends upon refractive index as $[n(n^2+2)^2/9]/n$, (Eq. (13)). Larger refractive index values are therefore expected to increase the stimulated emission cross-section.

The effective bandwidth of the emission spectra can be estimated using the relation:

$$\Delta\lambda_{eff} = \int \frac{I(\lambda)d\lambda}{I_{max}}$$ (18)

where $I(\lambda)$ is the measured fluorescence intensity at wavelength λ and I_{max} is the peak fluorescence intensity. Generally, the bandwidth is mainly caused by the splitting of the levels of different transitions and the site to site variation of the ligand field around Er^{3+} ions in glass, *i.e.*, the inhomogeneous broadening. The calculated values of stimulated emission cross-section and effective line width for fluorescence transitions $^4I_{13/2} \rightarrow {}^4I_{15/2}$ for all the glasses are listed in Table **5**. The stimulated emission cross-section of fluorescence transition depends upon intensity parameters in the following way:

➤ $^4I_{13/2} \rightarrow {}^4I_{15/2}$ depends upon Ω_2, Ω_4 and Ω_6 [more on Ω_6, and less on Ω_2, since $\|U^{(2)}\|^2$ have very small value].

The value of stimulated emission cross-section is highest for CBBE glasses and least for CSBE glasses. Generally, transitions with large stimulated emission cross-sections exhibit low threshold and high gain laser operation. An important parameter "Figure of Merit" (FOM) which defines the gain band-width of an amplifier [55] has been calculated and values are listed in Table **5** for all the glass samples. FOM is highest in CBBE glasses followed by ZBBE, ZSBE and CSBE glasses. Gain band-width of an amplifier tells us about the satisfactory performance of an amplifier over a particular frequency range. The performance may be different for different applications. The emission properties like FWHM, σ and FOM of Er^{3+} ions doped in different hosts are also shown in Table **5** for comparison. It is concluded that the fluorescence transition $^4I_{13/2} \rightarrow {}^4I_{15/2}$ in Er^{3+} ions is the potential lasing transition. The emission cross-section of a transition of interest can be enhanced by suitably selecting the host glass composition.

9. STRUCTURAL CHARACTERIZATION

9.1. Infrared Transmission Spectra

9.1.1. 20MO·xBi$_2$O$_3$·(79.5-x)B$_2$O$_3$·0.5Er$_2$O$_3$ (M = Zn and Cd) Glasses

Infrared spectroscopy is one of the most useful experimental techniques available for structural studies of glasses [56]. This technique leads to structural aspects related to both the local units constituting the glass network and the anionic sites hosting the modifying metal cations. The main vibrational modes appearing above 400 cm^{-1} in mid infrared range are associated with the structural change in the glass network [57-59]. These network modes are well separated from the metal ion site vibrational modes active in the far infrared region [59, 60]. The Fourier transform infrared (FTIR) spectrum for all Er^{3+} doped zinc/cadmium bismuth borate (ZBBE and CBBE) glasses were recorded and are shown in Figs. **11** and **12**, respectively.

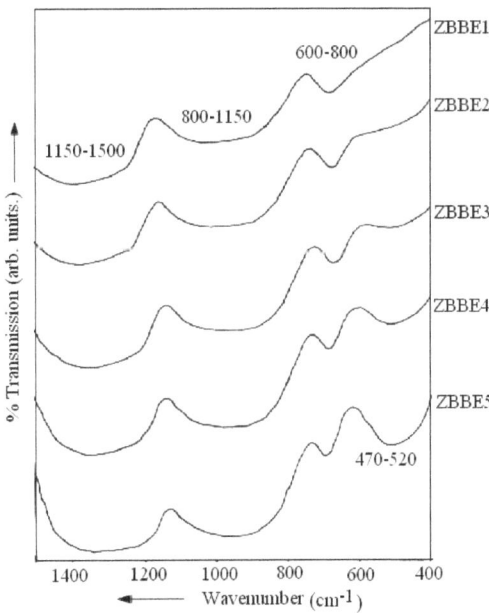

Figure 11: FTIR spectra of 20ZnO·xBi$_2$O$_3$·(79.5-x)B$_2$O$_3$ ·0.5Er$_2$O$_3$ (ZBBE) glasses.

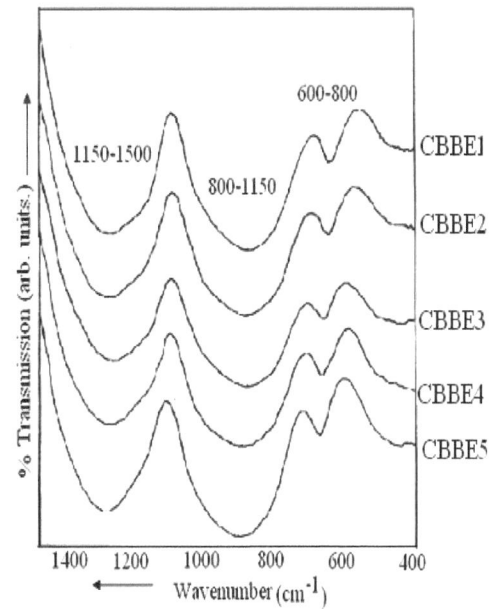

Figure 12: FTIR spectra of 20CdO·xBi$_2$O$_3$·(79.5-x)B$_2$O$_3$ ·0.5Er$_2$O$_3$ (CBBE) glasses.

The FTIR spectrum is given in the range of 400-1600 cm^{-1} for better clarity. It is observed that the IR spectra of these glasses arise largely from the modified borate networks. The structure of boron oxide glass consists of a random network of BO_3 triangles and BO_4 tetrahedral units. It is known that Bi_2O_3 can form (in glass networks) deformed BiO_6 units, both BiO_6 and BiO_3 units, and only BiO_3 pyramidal units [61]. The BiO_3 unit have four fundamental vibrations: a totally symmetric stretching vibration at 840 cm^{-1}, a doubly degenerate stretching vibration at 540–620 cm^{-1}, a totally symmetric bending vibration at 470 cm^{-1} and a doubly degenerate bending vibration at 350 cm^{-1} [53]. The vibrations of strongly distorted BiO_6 octahedral give the absorption bands centered at 860, 575–600, 470–520 and 430 cm^{-1} [62].

From Fig. **11**, it is observed that initially there is no band at 470-520 cm^{-1} but as the content of Bi_2O_3 increases the band starts arising and becomes deeper also. Therefore, it is proposed that the band from 470-520 cm^{-1} can be used to identify the BiO_6 units. Also, the vibration modes of the borate glasses occur in three IR spectral regions [63-66]: the first region 600-800 cm^{-1} is due to bending vibrations of various borate segments, the second region 800-1150 cm^{-1} can be attributed to the B-O stretching vibrations of BO_4 units and the third region 1150- 1550 cm^{-1} is due to the B-O and B-O-B stretching vibrations of BO_3 and BO_2O^- units. In the present glass system, the absence of absorption peak at 806 cm^{-1} indicates the absence of boroxol ring formation. Also, the absorption at 840 cm^{-1} is not observed in the glasses which suggest that the formation of tetrahedral coordination of Zn and Cd (*i.e.* MO_4 unit) is absent [67]. The low frequency band around 453 cm^{-1} (weakly observed) in the spectra of investigated glasses except in glass samples ZBBE4 and ZBBE5 may be attributed to the vibration of metal cations such as Zn^{2+} and Cd^{2+}. Since B_2O_3 is more acidic than ZnO or CdO, it takes up the oxide ions from Bi_2O_3 on priority for the modification of B–O–B bonds. Each BO_4 unit is linked to two such other units and one oxygen ion from each unit is linked with a metal ion and these linkages form long tetrahedron chains. It is assumed that the conversion of BO_3 (bridging) units to BO_4 (non-bridging) units takes place with increase in Bi_2O_3 concentration. This behavior is consistent with the variation of molar volume. Formation of BO_4 at the expense of BO_3 may increase the molar volume as glass structure becomes more open. This also indicates that the presence of Bi_2O_3 and ZnO or CdO in the glasses under study is as network former.

9.1.2. MO·xSiO₂·(79.5-x)Bi₂O₃·0.5Er₂O₃ (M = Zn and Cd) Glasses

The effect of substitution of unconventional glass former (Bi_2O_3) by the conventional glass former (SiO_2) on the structural properties of Er^{3+} doped zinc/cadmium bismuth silicate glasses was investigated by recording their Fourier transform infrared (FTIR) spectrum in the range of 400-4000 cm^{-1}. Figs. **13** and **14** show the FTIR spectra of all the ZSBE and CSBE glass samples in the range 400-1400 cm^{-1} for better clarity. Figs. **13** and **14** show a broad but strong band at 496-508 cm^{-1} in (ZSBE) and 488-496 cm^{-1} in (CSBE) and it shifts slightly towards lower wave number with decrease in Bi_2O_3/SiO_2 ratio.

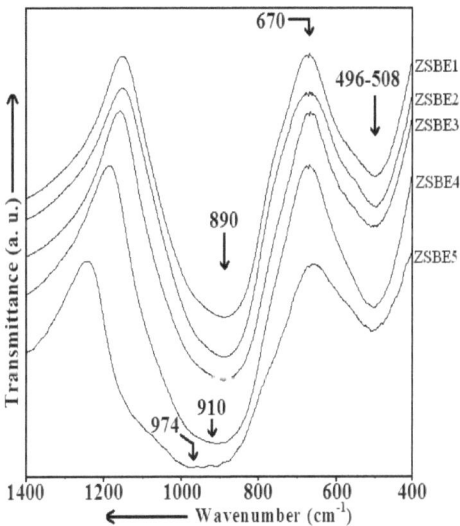

Figure 13: FTIR spectra of 20ZnO.xSiO2·(79.5-x)Bi2O3.0.5Er2O3 (ZSBE) glasses.

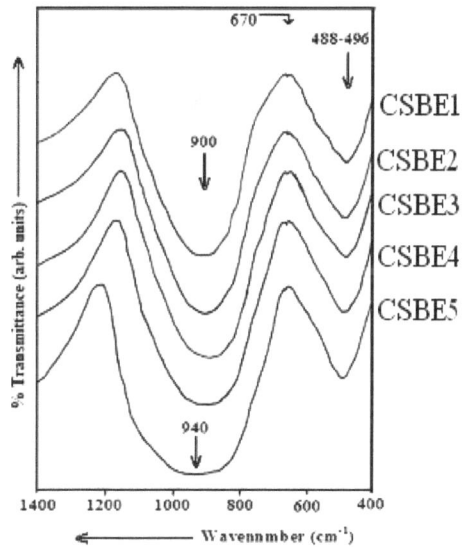

Figure 14: FTIR spectra of 20CdO·x SiO₂·(79.5-x)Bi₂O₃·0.5Er₂O₃ (CSBE) glasses.

It has been accepted by various authors [61, 68] that this band originates from Bi-O bonds in BiO_6 octahedra and the shifting to lower wave number is due to the decrease of the degree of distortion [69]. Dimitriev *et al.* [70] has reported that the shift in band around 482-520 cm^{-1} is attributed to the variation in the local symmetry of highly distorted BiO_6 polyhedra and the same was also observed in IR spectra of other bismuth based glasses [71]. Some authors [72-74] observed band in the wave number range 440-470 cm^{-1} and attributed to the symmetric oxygens bending-rock mode (R) BO's bonding. Thus the band near

496-508 cm^{-1} in ZSBE glasses and 488-496 cm^{-1} in CSBE glasses may compose of two bands but most of its intensity comes from bending vibrations of the silicate network as there is practically no change in the integrated intensity ratio between the band near 1000 cm^{-1} (stretching mode of silicate network) and the band near 496-508 cm^{-1} in ZSBE glasses and 488-496 cm^{-1} in CSBE glasses for the studied glasses in spite of a decrease in the concentration of Bi$_2$O$_3$. Therefore, shift of this band is due to depolymerization of the network and change in the Si-O-Si angle. The second prominent infrared band is observed around 800-1150 cm^{-1} centered at about 890 cm^{-1} (in ZSBE glasses) and 900 cm^{-1} (in CSBE glasses) for SiO$_2$ ≤ 30 mol% is due to Bi-O stretching of [BiO$_3$] pyramidal units [75] and is accompanied by a broad shoulder centered at around 1200 cm^{-1}. The band observed around 800-1150 cm^{-1} is linked to the stretching vibrations modes of the SiO$_4$ tetrahedra which is accompanied by a broad shoulder centered at around 1200 cm^{-1}. For higher amount of SiO$_2$, this band becomes broad and is accompanied by a shoulder centered at about 1200 cm^{-1} which is shifted to higher wave number. The blue shifting of this band with the successive replacement of Bi$_2$O$_3$ by SiO$_2$ suggests an increase in Si-O-Si bond angle [43]. In the spectrum of the present glass system, the broad band in the higher wave number range (800-1150 cm^{-1}) is linked to the stretching vibrations modes of the SiO$_4$ tetrahedra. A sharp peak at 670 cm^{-1} in all glasses (ZSBE and CSBE) was observed with increase in intensity as the Bi$_2$O$_3$ content decreases or SiO$_2$ increases. For the present investigation, the bands at 670, 620, 580 and 390 cm^{-1} in the spectrum of α-Bi$_2$O$_3$ were interpreted as vibrations of Bi-O bonds of different bond lengths in the distorted BiO$_6$ polyhedra. It was also found that no infrared band at 840 cm^{-1} corresponding to the [BiO$_3$] polyhedra [61] appeared (Figs. **13** and **14**). These experimental observations suggest that the [BiO$_3$] polyhedra are absent in the present glasses and therefore Bi^{3+} cations are interpreted in [BiO$_6$] groups only [61]. It was also found that there is no infrared band which corresponds to Si-O-Si stretched vibrations at 1120 cm^{-1} but a strong infrared band is observed at 1200 cm^{-1} in all glass compositions and the later band may be overlapped in this band. The structures of bismuth oxygen polyhedra are more flexible owing to the more ionic nature of the Bi-O bonds; therefore this network can incorporate the SiO$_4$ tetrahedra to a certain extent.

CONCLUSIONS

Optical properties, absorption and emission spectra of erbium doped zinc/cadmium bismuth borate/silicate glasses have been investigated. The refractive index of glasses increases from 1.75 to 1.95 for ZBBE and 1.71 to 1.91 for CBBE glasses with Bi_2O_3 content in the host, while in ZSBE and CSBE glasses it decreases from 2.19 to 2.06 and 2.22 to 2.13, respectively. From the absorption spectra, the Judd-Ofelt intensity parameters have been evaluated. The oscillator strength determined theoretically as well as experimentally is found to be in good agreement. The three intensity parameters Ω_λ (λ = 2, 4, 6) follows the trend: $\Omega_2 > \Omega_4 > \Omega_6$ for ZBBE and CBBE glasses while for ZSBE and CSBE the trend is $\Omega_4 > \Omega_6 > \Omega_2$, this may be due to change of host composition. From these intensity parameters and oscillator strength other radiative properties are also determined, *viz.*, radiative transition probabilities (A_{rad}), radiative lifetimes of the excited states (τ_r), stimulated emission cross-section (σ), figure of merit ($\sigma \times \Delta\lambda_{eff}$), total radiative transition probability (A_T) and total stimulated emission cross-section (σ_T). The effect of composition in erbium doped zinc/cadmium bismuth borate/silicate glasses on spectroscopic properties of the $^4I_{13/2} \rightarrow {}^4I_{15/2}$ transition, including the width of emission line and stimulated emission cross-section have been studied. Widths of emission lines are maximum for ZBBE3 (70 nm), CBBE1 (76 nm), ZSBE1 (70 nm) and CSBE1 (54 nm) glasses. The spectroscopic parameters of these glasses are better than other borate and silicate glasses and some of the spectroscopic parameters are comparable to the frequently used laser glasses. The red shift in the peak wavelength with Bi_2O_3 content in the fluorescence spectra is attributed to the nephlauxetic effect. The emission parameters, *viz.*, σ and FOM for emission at 1.5 μm have been compared with other hosts. The results of these investigations indicate that prepared Er^{3+} doped glasses may be useful for the development of 1.5 μm broadband amplifier. The analysis of IR transmission spectra shows that the basic network of present glasses consists of chains with $[BO_3]$, $[BO_4]$ and $[BiO_6]$ units for ZBBE and CBBE glasses. Also in case of ZSBE and CSBE glasses the band around 900 cm^{-1} is due to formation of $[BiO_3]$ units. There is formation of SiO_4 tetrahedra which accomplish a shoulder around 1200 cm^{-1} in ZSBE and CSBE glasses. A correlation between multi-phonon relaxation rates with the phonon energy of the network forming groups implies that the present glasses have high radiative quantum efficiency. The results of these investigations indicate that the spectroscopic properties Er^{3+}-doped zinc/cadmium bismuth borate/silicate glasses are better than bismuth borate and borate glasses, therefore these glasses may be useful for the development of lasers and fiber amplifiers.

CONFLICT OF INTEREST

There is no conflict of interest with other people or organizations in respect of the present research work.

ACKNOWLEDGEMENTS

The authors are thankful to UGC, CSIR and DST (FIST scheme) New Delhi, India for providing financial support.

REFERENCES

[1]   Wybourne BG. Spectroscopic Properties of Rare Earths. John Wiley & Sons Inc., New York 1965.

[2]   Dieke GH. Spectra and Energy Levels of Rare Earth Ions in Crystals. Crosswhite HM, Crosswhite H. Jhon Wiley & Sons Inc., New York 1968.

[3]   Dieke GH, Crosswhite HM. The Spectra of the Doubly and Triply Ionized Rare Earths. Applied Optics 1963; 2: 675-686.

[4]   Macfarlane RM, Shelby RM. Magnetic field dependent optical dephasing in LaF_3-Er^{3+}. Optics Communication 1982; 42: 346-350.

[5]   Ganem J, Jang KW, Boye D, Jones ML, Meltzer RS, Yen WM. Nonexponential photon-echo decays of paramagnetic ions in the superhyperfine limit. Physical Review Letter 1991; 66: 695-698.

[6]   Silberberg Y, Silva VL, Heritage JP, Chase EW, Andreco MJ. Accumulated photon echoes on doped fibers. IEEE Journal of Quantum Electronics 1992; 28: 2369-2381.

[7]   Jiangting S, Jiahua Z, Baojiu C. PMMA with Long-Persistent Phosphors and Its Behavior of Luminescence. Journal of Rare Earths 2004; 23(2): 157-159.

[8]   Huang YD, Mortier M Auzel F. Stark level analysis for Er^{3+}-doped ZBLAN glass. Optical Materials 2001; 17: 501-511.

[9]   Huang YD, Mortier M Auzel F. Stark levels analysis for Er^{3+}-doped oxide glasses: germanate and silicate. Optical Materials 2001; 15: 243-260.

[10]   Lin H, Pun EYB, Liu XR. Er^{3+}- doped $Na_2O \cdot Cd_3Al_2Si_3O_{12}$ glass for infrared and upconversion applications. Journal of Non-Crystalline Solids 2001; 283: 27-33.

[11]   Tanabe S, Sugimoto N, Ito S, Hanada T. Broad-band 1.5 μm emission of Er^{3+} ions in bismuth-based oxide glasses for potential WDM amplifier. Journal of Luminescence 2000; 87: 670-672.

[12]   Oprea II, Hesse H, Betzler K. Optical properties of bismuth borate glasses. Optical Materials 2004; 26: 235-237.

[13]   Saisudha MB, Rao KSRK, Bhat HL, Ramakrishna J. The fluorescence of Nd^{3+} in lead borate and bismuth borate glasses with large stimulated emission cross section. Journal of Applied Physics 1996; 80: 4845-4854.

[14]   Saisudha MB, Ramakrishna J. Optical absorption of Nd^{3+}, Sm^{3+} and Dy^{3+} in bismuth borate glasses with large radiative transition probabilities. Optical Materials 2002; 18: 403-417.

[15]   Judd BR. Optical Absorption Intensities of Rare-Earth Ions. Physical Review 1962; 127: 750-761.

[16]   Ofelt G S. Intensities of Crystal Spectra of Rare-Earth Ions. Journal of Chemical Physics 1962; 37: 511-521.

[17]   Becker P. Thermal and optical properties of glasses of the system Bi_2O_3 – B_2O_3. Crystal Research Technology 2003; 1: 74-82.

[18]   Chen Y, Huang Y, Huang M, Chen R, Luo Z. Spectroscopic properties of Er^{3+} ions in bismuth borate glasses. Optical Materials 2004; 25: 271-278.

[19]   Chen Y, Huang Y, Huang M, Chen R, Luo Z. Effect of Nd^{3+} on the Spectroscopic Properties of Bismuth Borate Glasses. Journal of American Ceramic Society 2005; 88: 19-23.

[20]   Yang JH, Dai SX, Zhon YF, Wen L, Hu LL, Jiang ZH. Broad-band 1.5 µm emission of Er^{3+} ions in bismuth-based oxide glasses for potential WDM amplifier. Journal of Applied Physics 2000; 87: 670-672.

[21]   Agarwal A, Pal I, Sanghi S, Aggarwal MP. Judd-Ofelt parameters and radiative properties of Sm^{3+} ion doped zinc bismuth borate glasses. Optical Materials 2009; 32: 339-344.

[22]   Pal I, Agarwal A, Sanghi S, Aggarwal MP. Structural, absorption and fluorescence spectral analysis of Pr^{3+} ions doped zinc bismuth borate glasses. Journal of Alloys and Compound 2011; 509: 7625-7631.

[23]   Sanghi S, Pal I, Agarwal A, Aggarwal MP. Effect of Bi_2O_3 on spectroscopic and structural properties of Er^{3+} doped cadmium bismuth borate glasses. Spectrochemica Acta A: Molecular and Biomolecular Spectroscopy 2011; 83: 94-99.

[24]   Paul MC, Harun SW, Huri NAD, *et al.* Performance comparison of Zr-based and Bi-based erbium-doped fiber amplifiers. Optics Letters 2010; 17,35: 2882-2884.

[25]   Carnall WT. The Absorption and Fluorescence Spectra of Rare Earth Ions in Solution, in: Handbook on the Physics and Chemistry of Rare Earths. Eds. Gschneidner KA. Jr, Eyring L. North-Holland Publishing Company 1979.

[26]   Reisfeld R, Jorgensen CK. Lasers and Excited States of Rare Earths. Springer-Verlag Berlin Heidelberg, New York, 1977.

[27]   Broer LJF, Gorter CJ, Hoogschagen. On the intensities and the multipole character in the spectra of the rare earth ions. Physica 1945; 11: 231-250.

[28]   Brown DC. High Peak Power Nd: Glass Laser Systems. Springer-Verlag Berlin Heidelberg, New York, 1981.

[29]   Krupke WF. Induced-emission cross sections in neodymium laser glasses. IEEE Journal of Quantum Electronics 1974; QE-10: 450-457.

[30]   Krupke WF. Optical Absorption and Fluorescence Intensities in Several Rare-Earth-Doped Y_2O_3 and LaF_3 Single Crystals. Physical Review 1966; 145: 325-337.

[31]   Carnall WT, Fields PR, Rajnak K. Spectral Intensities of the Trivalent Lanthanides and Actinides in Solution. II. Pm^{3+}, Sm^{3+}, Eu^{3+}, Gd^{3+}, Tb^{3+}, Dy^{3+}, and Ho^{3+}. Journal of Chemical Physics 1968; 49: 4412-4424.

[32] Carnall WT, Fields PR, Wybourne BG. Spectral Intensities of the Trivalent Lanthanides and Actinides in Solution. I. Pr^{3+}, Nd^{3+}, Er^{3+}, Tm^{3+}, and Yb^{3+}. Journal of Chemical Physics 1965; 42: 3797-3807.

[33] Sarkies PH, Sandoe JN, Parke S. British Journal of Applied Physics (Journal Physics) 1971; 4: 1642-1650.

[34] Keil A. Multiphonon Spontaneous Emission in Paramagnetic Crystals, in: Quantum Electronics, Eds. Grivet P. and Bloemberger N. New York, Columbia, UP, 1964.

[35] Jorgensen CK, Judd BR. Molecular Physics 1966, 8: 281-290.

[36] Judd BR. Hypersensitive Transitions in Rare-Earth Ions. Journal of Chemical Physics 1966; 44: 839-841.

[37] Peacock RD. Structure and Bonding 1975; 22: 83.

[38] Weber MJ. Rare Earth Lasers, in: Handbook on the Physics and Chemistry of Rare Earths. Eds. Gschneidner KA. Jr, Eying L. North-Holland Publishing Company 1979.

[39] Drake CF, Stephan JA, Yates B. The densities of V_2O_5/P_2O_5 glasses and the oxygen molar volume. Journal of Non-Crystalline Solids 1978; 28: 61-65.

[40] Shankar S, Dasgupta A, Basu B, Paul A. Journal of Materials Science Letters 1983; 4: 697-702.

[41] Shannon R. Revised effective ionic radii and systematic studies of interatomic distances in halides and chalcogenides. Acta Crystallor A 1976; 32: 751-767.

[42] Wells A. Structural Inorganic Chemistry. Fourth Ed., Clarendon Press, Oxford, 1975.

[43] Pan Z, Henderson DO, Morgan SH. Vibrational spectra of bismuth silicate glasses and hydrogen-induced reduction effects. Journal of Non-Crystalline Solids 1994; 171: 134-140.

[44] Morsi MM, El-Konsol S, Adawi MA. Effect of neutron and gamma irradiation on some properties of borate glasses. Journal of Non-Crystalline Solids 1983; 58: 187-199.

[45] Krupke WF. Radiative transition probabilities within the 4f ^{3}ground configuration of Nd:YAG. IEEE Journal of Quantum Electronics 1971; QE-7: 153-159.

[46] Lucas J, Chanthanasinh M, Poulain M, Burn P. Preparation and optical properties of neodymium fluorozirconate glasses. Journal of Non-Crystalline Solids 1978; 27: 273-283.

[47] Weber MJ, Saroyan RA, Ropp RC. Optical properties of Nd^{3+} in metaphosphate glasses. Journal of Non-Crystalline Solids 1981; 44: 137-148.

[48] Gatterer K, Pucker G, Fritzer HP, Arafa S. Hypersensitivity and nephelauxetic effect of Nd(III) in sodium borate glasses. Journal of Non-Crystalline Solids 1994; 176: 237-246.

[49] Izumitani T, Toratani H, Kuroda H. Radiative and nonradiative properties of neodymium doped silicate and phosphate glasses. Journal of Non-Crystalline Solids 1982; 47: 87-99.

[50] Nageno Y, Takebe H, Morinaga K. Correlation between Radiative Transition Probabilities of Nd^{3+} and Composition in Silicate, Borate, and Phosphate Glasses. Journal of American Ceramic Society 1993; 76: 3081-3086.

[51] Oprea II, Hesse H, Betzler K. Luminescence of erbium-doped bismuth–borate glasses. Optical Materials 2006; 28: 1136-1142.

[52] Canalejo M, Cases R, Alcala R. Optical properties of Sm^{3+} in fluorozirconate glasses. Physics and Chemistry of Glasses 1988; 29: 187-194.

[53] Weber MJ. Fluorescence and glass lasers. Journal of Non-Crystalline Solids 1982; 47: 117-133.

[54] Deutchbein OK, Pautrat CC. CW laser at room temperature using vitreous substances. IEEE Journal Quantum Electronics 1968; QE-4: 48-51.

[55] Moorthy LR, Jayasimhadri M, Saleem SA, Murthy DVR. Optical properties of Er^{3+}-doped alkali fluorophosphate glasses. Journal of Non-Crystalline Solids 2007; 353: 1392-1396.

[56] Wong J, Angell CA. Glass Structure by Spectroscopy. Marcel Dekker Inc. New York, 1967.

[57] Witkowska A, Regbicki J, Eicco AD. Structure of partially reduced bismuth–silicate glasses: EXAFS and MD study. Journal of Alloys and Compounds 2005; 401: 135-144.

[58] Bradry KMEl, Moustaffa FA, Azooz MA, Batal FHEl. Indian Journal Pure Applied Physics 2000; 38: 741-745.

[59] Krogh-Moe J. Interpretation of the infra-red spectra of boron oxide and alkali borate glasses. Physics and Chemistry of Glasses 1965; 6: 46-58.

[60] Merzbacher CI, White WB. The structure of alkaline earth aluminosilicate glasses as determined by vibrational spectroscopy. Journal of Non-Crystalline Solids 1991; 130: 18-34.

[61] Baia L, Stefan R, Popp J, Simon S, Kiefer W. Vibrational spectroscopy of highly iron doped B_2O_3–Bi_2O_3 glass systems. Journal of Non-Crystalline Solids 2003; 324: 109-117.

[62] Hu Y, Liu NH, Lin UL. Journal of Materials Science 1998; 33: 229.

[63] Yiannopoulos YD, Chryssikos GD, Kamitsos EI. Structure and properties of alkaline earth borate glasses. Physics and Chemistry of Glasses 2001; 42: 164-172.

[64] Kamitsos EI, Karakassides MA, Chyssikos GD. Far-infrared spectra of magnesium-sodium-borate glasses. Solid State Communications 1986; 60: 885-888.

[65] Abo-Naf M, Batal FHEl, Azooz MA. Characterization of some glasses in the system SiO_2, $Na_2O \cdot RO$ by infrared spectroscopy. Materials Chemistry and Physics 2002; 77: 846-852.

[66] Kumar A, Rai SB, Rai DK. Effect of thermal neutron irradiation on Gd^{3+} ions doped in oxyfluoroborate glass: an infra-red study. Materials Research Bulletin 2003 ; 38: 333-339.

[67] Motke SG, Yawale SP, Yawale SS. Bulletin of Materials Science 2002; 25: 75-82.

[68] Bray PJ, O'Keefe JG. Nuclear magnetic resonance investigations of the structure of alkali borate glasses. Physics and Chemistry of Glasses 1963; 4: 37-47.

[69] Griscom DL. Borate Glasses in: Materials Science Research, vol. 12, Eds. Pye LD, Frechate VD, Kriedle NJ. Plenum Press, New York, 1978.

[70] Dimitriev Y, Mihailova M. Proceedings of the 16th International Congress on Glass. Volume 3, Madrid, 1992.

[71] Bale S, Purnima M, Srinivasu C, Rahman S. Vibrational spectra and structure of bismuth based quaternary glasses. Journal of Alloys and Compound 2008; 457: 545-548.

[72] Bell J, Dean P. Discussion Faraday Society 1970; 50: 55-67.

[73] Sen PN, Thorpe MF. Phonons in AX_2 glasses: From molecular to band-like modes. Physical Review B 1977; 15: 4030-4038.

[74] Galeener FL. Band limits and the vibrational spectra of tetrahedral glasses. Physical Review B 1979; 19: 4292-4297.

[75] Sreenivasu D, Chandramouli V. EPR, IR and DC conductivity studies of xCuo-(100 -x) B_2O_3 glasses. Bulletin of Materials Science 2000; 23: 509-516.

Keyword Index

A

Al$_2$O$_3$ 3, 9, 20, 26, 30

B

Bioactive   3, 6, 10, 13, 52
Bioceramics   3, 8, 27, 41, 73
Biocompatibility   3, 10, 49, 87, 104
Bioglasses   10, 11
Biological characterization   87, 90, 91, 93
Biomaterials   3, 4, 41, 52, 88
Bone grafts   116, 119, 122, 135, 136
Bone regeneration   90, 108, 116, 122, 126
Bone tissue   87, 103, 110, 119

C

Cadmium bismuth borate   142, 146, 170, 173, 177
Calcium   3, 22, 89, 103, 116
Cell adhesion   87, 104, 107, 108, 110
Cell culture   87, 103
Cell proliferation   104, 107, 109
Ceramics   3, 10, 27, 35, 38
Cerium oxide   87
Commercial ceramics   49
Composites   8, 12, 57, 73, 87
Conventional heating   73, 74, 78
Cytocompatibility   87, 110

D

Dentistry   9, 49, 52, 57, 64
Doped glasses   87, 90, 142
Doped hydroxyapatite composites   87

Sooraj H. Nandyala and José D. Santos (Eds)
All rights reserved-© 2013 Bentham Science Publishers

Author Index

Sooraj H. Nandyala and José D. Santos (Eds)
All rights reserved-© 2013 Bentham Science Publishers

S

www.ingramcontent.com/pod-product-compliance
Lightning Source LLC
Chambersburg PA
CBHW041701210326
41598CB00007B/493